Geomorphology of Central America

A Syngenetic Perspective

Tárcoles River Mouth in Costa Rica

Geomorphology of Central America

A Syngenetic Perspective

Jean Pierre Bergoeing

ELSEVIER

AMSTERDAM • BOSTON • HEIDELBERG • LONDON • NEW YORK • OXFORD
PARIS • SAN DIEGO • SAN FRANCISCO • SINGAPORE • SYDNEY • TOKYO

Elsevier
Radarweg 29, PO Box 211, 1000 AE Amsterdam, Netherlands
The Boulevard, Langford Lane, Kidlington, Oxford OX5 1GB, UK
225 Wyman Street, Waltham, MA 02451, USA

ISBN: 978-0-12-803159-9

British Library Cataloguing in Publication Data
A catalogue record for this book is available from the British Library

Library of Congress Cataloging-in-Publication Data
A catalog record for this book is available from the Library of Congress

For information on all Elsevier publications
visit our website at http://store.elsevier.com/

Working together
to grow libraries in
developing countries

www.elsevier.com • www.bookaid.org

Contents

About the Author

Jean Pierre Bergoeing is a French geomorphologist who studied at the Pontifical Catholic University of Chile. He pursued further study at the University of Aix-Marseille II, France, where he received a master's degree in physical geography (1972), a third-cycle doctorate in geomorphology (1975), and a state letters and human sciences doctorate (1987). His career has developed across three continents: America, Europe, and Africa. He has been a professor at the Pontifical Catholic University of Chile; the University of Costa Rica; the University of Nantes, France; and Abdou Moumouni University of Niamey, Niger. He has also served in a diplomatic position for the government of France as an international cooperator, and later as a diplomat, with the title of Scientific and Technical Cooperation Attache. He has authored numerous publications in international journals and geomorphological maps of Chile, Costa Rica, Central America, Africa, and Europe, and has served as a professor at the University of Costa Rica since 2005.

Introduction

Poas crater volcano, Costa Rica. Aerial photography taken by J.P. Bergoeing, 2012.

From a geomorphological point of view, Central America is characterized by a series of original landscapes or regions resulting from a slow but vigorous tectonic and volcanic evolution throughout the geological ages. It is thus that we distinguish two major areas in its formation:

1. The nucleus Central America ranges from the Tehuantepec isthmus to Northern Nicaragua and has a long geological history, characterized by different orogenic processes (orogenic cycles of the Precambrian, Hercynian, and Laramian, which cover from the Jurassic until today and have been characterized by great intensity). These processes characterize the current landscape, with sedimentary, metamorphic, and intrusive reliefs dating back to the Primary, Secondary, and Tertiary periods.
2. The Isthmus Central America is much younger and is distributed from Northern Nicaragua to Panama; its reliefs only began to emerge in the late Cretaceous period, about 80 million years ago.

It can be said that Central America is the result of the clash of two minor tectonic plates (e.g., Cocos plate, covering part of the Pacific floor, and the

Caribbean plate) colliding against each other about five million years ago. This gave rise to several vigorous reliefs by effect of the orogenesis associated with magma rising from the mantle through the Earth's crust, weakened by the collision of the aforementioned tectonic plates. This weakness allowed the ascension of the red-hot magma through fissures that fed magmatic chambers hosted in the Earth's crust. This in turn gave rise to new volcanoes, which in some cases were completed by thunderous eruptions that made existing volcanic structures collapse, leaving vast depressed areas known as calderas. For this event we can mention varied examples, such as the Managua caldera in Nicaragua, Ilopango caldera in El Salvador, or Atitlan caldera in Guatemala. In areas farther away from the subduction zone, the constant sedimentation, landslides, and general erosion of the emerging volcanic ranges resulted in vast flooded plains that gradually gained dominion over the prevailing sea, obstructing contact between the Pacific Ocean and the Caribbean Sea, creating an isthmus.

Thus, since the beginning of the Quaternary period, vast landscapes have contrasted between coastal plains forming and new volcanic ridges building, being rapidly colonized by the tropical vegetation of those low latitudes, and then crossed by rivers whose headwaters loomed increasingly by effect of the orogeny, reaching altitudes higher than 9800 ft. Swift and plentiful, rivers run until reaching the plains where the slope breaks, making them slow and lazy but nevertheless plentiful courses. They feature countless meanders before reaching the sea, forming estuaries in some cases and deltas in the great majority, which go to the sea, feeding coastal strips due to littoral drift currents, in some cases.

In the high peaks, above 9800 ft, the area experienced two glacial periods associated with the large Quaternary climate changes suffered by the planet, especially in Alto Cuchumatanes (Guatemala) and Chirripo (Costa Rica). It's in this context, during the Holocene and probably in the Upper Pleistocene, that early human groups emigrated from distant lands and settled in these new latitudes, creating splendid civilizations such as the Maya, Chorotega, and Diquis cultures.

CHAPTER 1

Structural Geomorphology

TECTONIC PLATES

In 1915, German astronomer and meteorologist **Alfred Wegener** (1880-1930) published his theory of the *"Genesis of the continents and oceans,"* associating geophysics, geology, and geography. He revolutionized the concept of the cool Earth's crust. His theory lacked the concept of magma convection and attributed the continental drift to moon tides, an unsustainable element that was ridiculed by English geologist **Harold Jeffreys** and was the object of ridicule and acidic criticism from the scientific community. Fortunately, Wegener found support in other scientists, such as Swiss **Emile Argand**, Scotsman **Arthur Holmes**, and American **Reginald Daly**, who appreciated his scientific contribution to Earth sciences, which is today recognized worldwide. From the mid-twentieth century, advances in geophysical knowledge and particularly paleomagnetic studies have brought new knowledge and evidence of continental drift that allowed the reconstitution of the continents' position 200 million years ago. In 1923, Dutchman **Vening-Meinesz** discovered important gravimetric anomalies in the Indonesian pit with measurements taken from a submarine. This allowed him to determine that the terrestrial crust in that place had important large-scale flexures with elastic properties. We should expect, however, that the discoveries of **Hess** in 1960 on the seabed expansion would allow the imposing of the mobilistic current and the formulating of plate tectonics theory, instituted prematurely by Wegener. The hypothesis allowed for much more determination later on that the Earth's crust was split into a huge puzzle formed by plate tectonics. These plates are in constant movement, and the current position of the continents is the result of this migration of the Earth's crust sliding over the asthenosphere, where magma suffers converging or Brownian movements according to this hypothesis. Tectonic plates are constrained by large fractures that can be divided into three categories (Figure 1.1):

a. Accretion or dorsal areas
b. Landslide areas by transforming faults
c. Subduction zones

CONTENTS

1

Geomorphology of Central America. http://dx.doi.org/10.1016/B978-0-12-803159-9.00001-7

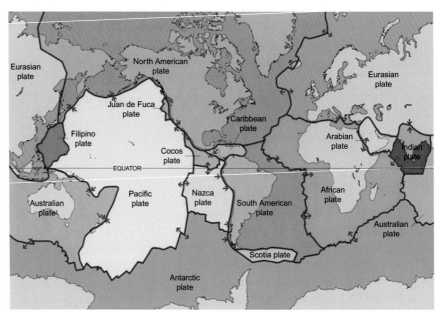

FIGURE 1.1
Tectonic plates, global distribution (http://www.bucknell.edu/x17758.xml).

Accretion or Dorsal Areas

Accretion or dorsal areas correspond to divergent tectonic plates areas of the Earth's crust. Separation produces a vacuum that is immediately filled by magmatic ascent. This activity usually produces basaltic lava flows in a marine environment, because the magma is not mixed with acidic rocks of the Earth's crust. Accretion zones generally correspond to middle ocean ridges, like the Atlantic Ridge, which corresponds to a fracture in the Earth's crust that runs from the Arctic Circle to the Antarctic Circle. A Quaternary-emerged testimony of magmatic ascent, products of the separation of both Atlantic plates, which in this sector reach a distance of 3/64 in. per year, are the islands and archipelagos in the Atlantic (e.g., Iceland, Azores, Tristan de Cunha). The great African Rift is also an accretion area that goes from Sinai to Mozambique. It's a great accident of stepped faults (graben), where important volcanic cones emerge as the Kilimanjaro, Meru, or Niragongo. It houses lakes that reach depths of more than 5577 ft. The Great Rift diverges the African plate from Somalian plate and will separate them in a few million years. The created space will be occupied by a new sea, as happened during the Jurassic era with Madagascar Island or with the Red Sea (Figure 1.2).

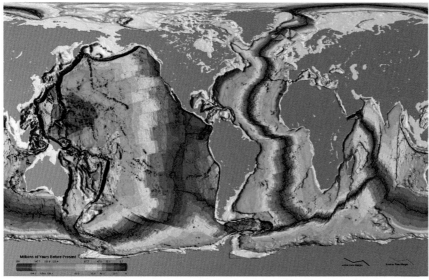

FIGURE 1.2
Mid-ocean ridge. Accretion area with tectonic plates diverging and separating 3 in./year, depending on the sector. Convection currents in the asthenosphere produce plate divergence and allow magmatic ascent in the fractured zone (www.geologycafe.com).

Landslide Areas by Transforming Faults

The second types of plate tectonic contact are those that do not come into direct collision, but rather slide in opposite directions. They transform a trench in a ridge, a dorsal in another dorsal, or a trench in another trench (e.g., the boundary between tectonic plates in the Caribbean and North America along the Chixoi-Polochixc-Motagua fault system). The boundary of the Juan de Fuca plate in California with the North American plate is another example of this second transforming fault system. Earthquakes of high intensity occur equally each time the plates are arranged as they move. These movements are able to cause considerable superficial changes and affect human life. Efforts accumulated between two plates release important energy that takes the form of earthquakes that can be catastrophic, such as the San Francisco, California, earthquake of 1906, due to the San Andreas transforming fault (Figure 1.3).

Subduction Zones

Subduction zones are confrontation areas of two or more tectonic plates colliding on an ongoing basis. This collision makes one plate go under the other, producing a leaning area of seismicity known as a Wadati-Benioff zone, which dives, in some cases, up to 434 mi to the interior of the Earth with an

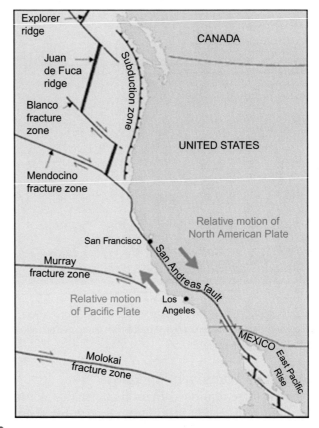

FIGURE 1.3

Transforming faults in California. The Mendocino's transforming fault (TF) allows the ocean bottom generated by the Juan de Fuca dorsal to move southeast with respect to the Pacific plate and south of the North American plate. Thus, this transforming fault connects a divergent zone (the ridge) to a subduction zone. In addition, the San Andreas Fault, also a TF, connects two centers of drift: Juan de Fuca's and a divergent ridge zone that exists in the Gulf of California (www.platetectonic.com).

inclination of 40°-60°, while the other plate's ascent gives rise to mountainous reliefs. Subducted plates are generally denser, consist mainly of gabbros and peridotite, and descend gradually toward the interior of the Earth and end up melting in the mantle with the underlying magma. In the subduction process, when a plate reaches depths of 260,000-330,000 ft, hydrated minerals cease to be stable at these depths and temperatures, and they move on to more stable structures and release water they contained (Figure 1.4).

That liberated water reduces the melting point of materials of the mantle and melts them partially. From there, the magmatic ascents infiltrating through the

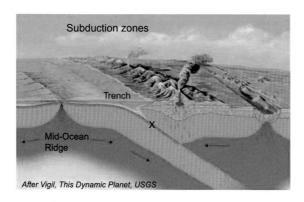

FIGURE 1.4

Vertical cut of two tectonic plates' collision and subduction area (x). Colliding plates produce the descent of the subducted plate into the asthenosphere, where it dehydrates, melting partially. Water released also partially melts the mantle. The ascending plate is followed by volcanic building, fed by the magmatic chambers' product of magma ascents (www.biologyeducation.net).

collision zone are magmatic reservoirs in the cortex that feed volcanoes that are the external expression of the magmatic surge. These bags or magmatic reservoirs are located approximately 20,000-30000 ft deep in the Earth's crust. It's the difference in density between the oceanic lithosphere and the asthenosphere that increases with time; the latter grows less rapidly, creating the real engine of the subduction zone. The oceanic lithosphere, heavier by increasing density with age, acquires the tendency to immerse itself. Subduction zones, also known as active margins, are where earthquakes and volcanic activity are most frequent (e.g., the Pacific Ocean's Ring of Fire or the Indonesian southern margin).

VOLCANISM

Volcanism as a Product of Tectonic Plates

Volcanic eruptions vary by magmatic acidity ascents that are defined by the percentage of silica (SiO_2) content in lavas. Basic basalts stood at 52%, followed by andesitic, dacitic, and rhyolitic lavas, with the last one reaching 58% and being the more acidic lava. Thus, the Hawaiian volcanic eruptions are predominantly basaltic, and are characterized by fluid lava flows that can reach 12-18 mi away. In order, they are Strombolian, Vulcanian, and Pelean eruptions. The last ones are the most acidic and explosive, and are characterized by fiery clouds or pyroclastic flows (*nuees ardentes*). The fiery cloud is a gaseous, viscous magma that reaches up to 1000 °C and is so dense that it descends the slopes of the volcanic cone, characterized by a large amount of pumice. Its deposits can form important thickness of ignimbrites, which are by nature very porous rocks. Thus was formed the structural plateau of Liberia in Costa Rica, whose vegetation denotes the difficulty to thrive despite the tropical damp climate.

Volcanism in the island arcs or in continental subduction areas occurs by dehydration of the subducted plate once it reaches more than 49 mi in depth. Water released goes to the mantle and lowers the melting point of materials, which are then found dry in the solid state, and partially fused. The molten material is lighter than the adjacent one and therefore ascends to the surface, forming volcanic mountain chains parallel to the subduction zone.

Hot Spot Volcanism

Besides subduction zone volcanism, also known as arc volcanism and mid-ocean ridge volcanism, another type of volcanic activity is present on our planet, which is not associated with boundaries of plates. Hot spot volcanism occurs by the rise of magma from deep levels in the mantle with a great deal of heat flux. This volcanism generates volcanic chains with activity at one end, where the volcanoes are progressively older toward the other end of the volcanic chain. This occurs by the slide of a plate, either oceanic or continental, on the hot point, which leaves behind a scar on that plate. Classic hot spot examples are the Hawaiian Islands and the submerged Emperor chain in the Pacific Ocean or the Galapagos Islands. As a hot spot on a ridge point, it leaves its scar on two plates: the Cocos and the Nazca plate. Cocos Island is located on the submerged chain of the Cocos plate, which is the scar left by the hot spot on the plate of the same name. A continental case is Yellowstone in the United States (Figure 1.5).

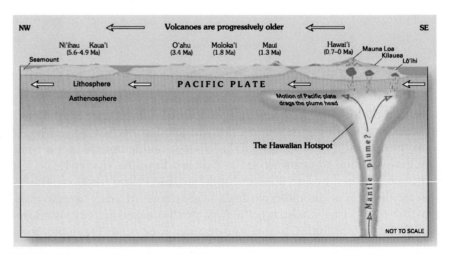

FIGURE 1.5
Hot spot volcanism in the Hawaiian archipelago (Wikipedia.org).

THE CENTRAL AMERICAN CASE

Regional tectonics of Central America are mainly controlled by the collision of the Cocos and the Caribbean plates. As a result of this collision, the oceanic Cocos plate is subducted below the Caribbean along the Mesoamerican trench plate at speeds ranging from 2 in. per year, compared to a little more than 3 in. per year in Guatemala, compared to the Osa Peninsula in Costa Rica (Protti-Quesada, 1994, calculated on the basis given by DeMets et al., 1990). It is along this plate boundary that most of the earthquakes of great magnitude occur in Costa Rica. At the southwest boundary of the Caribbean plate, the local condition of tectonic forces resulted in the fracturing of the same plate and the creation of a so-called microplate block of Panama, whose boundaries are not yet well developed nor defined. The northern boundary of the Panama block with the Caribbean plate is a convergence margin known as "Northern Panama deformed block" (Silver et al., 1990), which extends from the coast of the Caribbean from Colombia to Limon in Costa Rica. It was at the end of this plate boundary that the Estrella Valley earthquake occurred in April 1991. Toward the northwest, the contact between the Panama block and the Caribbean plate consists of a diffuse faulting zone of left-side shift, which runs from Limon to the Mesoamerican trench through the central part of Costa Rica (Jacob and Pacheco, 1991; Güendel and Pacheco, 1992, Goes et al., 1991, Ponce and Case, 1987; Fisher et al., 1994; and Protti and Schwartz, 1994). The Panama block comprises the southern part of Costa Rica and all of Panama (Figure 1.6).

The Panama fracture zone is located south of Burica Peninsula. This right-shifting fault system constitutes the limit of transformation between the Cocos and Nazca plates (cf. Figure 1.1). The Cocos underwater mountain range is located west of the Panama fracture zone, which is subducted under the Osa Peninsula. The nonseismic Cocos range is the scar formed on the Cocos plate as a result of its passage over the hot spot of the Galapagos Islands. However, the Landsat 2000 radar satellite images have allowed the highlighting of a contact zone between Costa Rica and Panama, by a wide variety of faults, that opens from the Central Valley to the coastal region between Barranca River and Quepos Peninsula.

SEISMOLOGY

Seismology is a discipline, a branch of geophysics for the study of earthquakes and the propagation of seismic waves that emerge from such movements. Earthquakes are caused by the accumulation of tensions, before displacement or adjustment of tectonic plates, and the interaction between them. There are also earthquakes generated by volcanism, but these are local and therefore

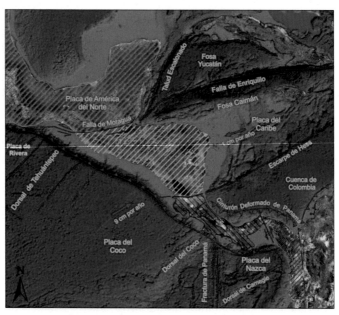

FIGURE 1.6

Central America's tectonic blocks and plates. 1—tectonic plate movement sense; 2—reversed or thrust fault; 3—transforming faults. Tectonic blocks: 4—North American plate; 5—Chortis block; 6—Chorotega block; 7—Choco block. *Photo interpretation by J.P. Bergoeing 2012, using 2012 satellite Google image and 2001 CEPREDENAC information source.*

smaller in the regional dimension. Earthquakes are studied using electronic seismographs that allow measuring of the propagation of the seismic waves. Primary (P) waves and secondary (S) waves originated in place at a depth where the earthquake (hypocenter) occurs, and surface waves (Love (L) and Raleigh (R)) are a product of the interaction of the waves with the Earth's surface. The epicenter is the projection to the surface, perpendicular to the hypocenter that reflects the intensity of an earthquake, a product of the liberation of tensions in the failure or weakness area in the Earth's crust. P waves are the faster ones, moving from 3 mi/s to 6 mi/s. They are the first to be registered. Their speed depends on the traversed material, which generally increases with depth. Move to the inside of the globe, and they can pass through the nucleus. S waves are waves in shear, vibrating perpendicular to the direction of the displacement of the wave; they cannot pass through the outer core for behaving like a liquid. Finally, L and R waves are the slowest, moving in the Earth's surface about 2 mi/s.

Giuseppe Mercalli (1850-1914) was an Italian volcanologist who devised a modern scale of seismic measurements based on the intensity of the tremors and damage caused by them. The scale is distributed from 1 to 12 and replaced the ancient Rossi-Forel scale. However, it was American seismologist Charles Richter (1900-1985) who developed the current scale of seismic magnitude or extent of the energy released by an earthquake at its source. Due to problems of saturation of instruments, this scale may not exceed 8 degrees. According to this scale, the magnitude of an earthquake (M) can go from 1 to 8 based on the time elapsed between the occurrence of P and S (Δt) waves and amplitude (A) waves, which correspond to the logarithm of the amplitude of the vibrations recorded by a seismograph measured as a function of the epicentral distance. This can be expressed by the following formula: $M = \log 10\, A\,(\text{mm}) + 3\,\log 10\,(8\,\Delta t(s)) - 2.92$ (Figure 1.7).

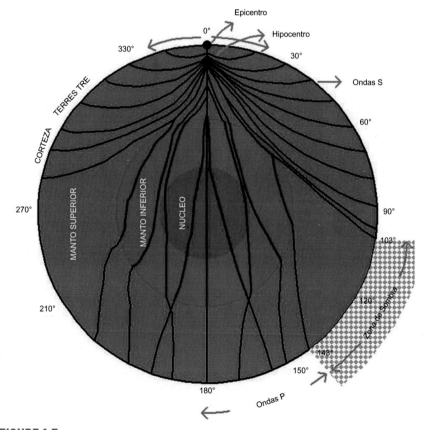

FIGURE 1.7
When an earthquake occurs, primary (P) and secondary (S) displacement waves move to the interior of the planet, and love (L) waves move on the crust. *Source: Department of Geography, University of Costa Rica, 2012.*

Given the saturation problem of the Richter scale, the scale of magnitude of seismic moment (M_w) is now used. This can be obtained at long distances with instruments of broadband or long period that are not saturated at these distances. The 1960 Chilean earthquake had 9.5 magnitude M_w, and the Alaskan earthquake had 9.2 M_w. On December 26, 2004, an earthquake shook the Banda Aceh sector northwest of Indonesia. The epicenter was located in Simeulue island, at the edge of the subduction arch of the Malay Archipelago. The earthquake had a magnitude $M_w = 9.3$ and was the second most intense ever recorded by a seismograph on the Richter scale, which caused a vibration of 03/64 in. of Earth's axis. By its magnitude, the surface of the source, and its offshore location, the earthquake caused a tsunami (tidal wave), which affected the coasts of the Indian Ocean, reaching up the Kenyan coast and particularly the Bay of Bengal. The tsunami and earthquake caused more than 300,000 fatalities (Figure 1.8).

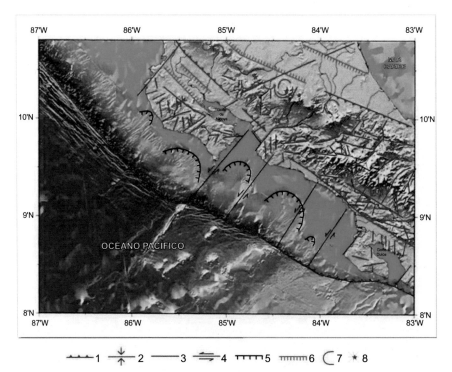

FIGURE 1.8

Collision zone of Cocos and Caribbean plates in Costa Rica with strike-slip fault system and large areas of submarine landslides on the continental slope generator of tsunamis. 1—Reverse faults; 2—Synclines; 3—Normal faults; 4—Strike-slip faults; 5—Submarine landslides; 7—Graben Edge; 8—Calderas; 9—Volcanic craters. *Photo interpretation by J.P. Bergoeing, 2012. Source: Geomar no.94, 1999.*

In Costa Rica, the largest earthquakes typically occur by effects of the subduction of the Cocos plate under the Caribbean plate and the Panama block, and in synergy with the Nazca plate. That is why the hottest places are situated in the Nicoya and Osa Peninsulas and the Southern Caribbean. Using high-resolution seismic data collected by the OVSICORI (Volcano Observatory and Seismology of Costa Rica), Protti et al. (1996) could determine the geometry of the subduction zone or Wadati-Benioff zone in Southern Central America. This data of 9500 earthquakes was used with a margin of error of less than 3 mi for the hypocentral location. This study allowed the determination that the Wadati-Benioff zone is much more tilted in Nicaragua than in Costa Rica, without faulting evidence. Further south, in Costa Rica, Protti et al. (1996) described the "sharp contortion of Quesada" area, which is located in the sector under Ciudad Quesada at a depth of 70,000 ft, and where the angle of inclination of the Wadati-Benioff zone decreases from 84° to 60° in the central part of Costa Rica. Similarly, the seismic hypocenters decrease in depth from 790,000 ft in Nicaragua to 410,000 ft under the central volcanic mountain range of Costa Rica. It is determined that the subduction of the Cocos plate under the southern region of Costa Rica is less inclined, and is responsible at the same time for the strong raises of the tectonic external arch.

Federico Güendell, PhD, of the OVSICORI (in the case study of the April 22, 1991 Limon earthquake) identified three strong seismic activity cycles in Costa Rica during the twentieth century, characterized by a great earthquake centered in the southeast of Costa Rica due to subduction, followed by other events of lesser magnitude due to interplate faults activated by the initial event. The first cycle was completed by an earthquake of great magnitude in Northern Costa Rica. According to Güendell, the first cycle began in 1904 with an earthquake of 7.5 whose epicenter was located in Golfo Dulce, and ended with the earthquake of 1916 on the northern border. To this first cycle belong the 1905 earthquake of Puntarenas and the 1910 earthquake of Cartago with a magnitude of 6. The second cycle began in 1914 with an earthquake of 7.5 whose epicenter was again located in Golfo Dulce, and ended in 1950 with the Nicoya earthquake of 7.7. Finally, the third cycle started in 1983 with an earthquake of 7.3 with its epicenter in Golfito (Golfo Dulce), followed by the earthquakes of Cobano (Nicoya), Alajuela (1990), and Limon in 1991.

In the twenty-first century, an earthquake of magnitude 6.4 occurred on November 20, 2004, with its epicenter in the Central Pacific of Costa Rica at a depth of 16 mi, and 62 mi from the Mesoamerican trench studied by the OVSICORI. Applying the technique of the double difference by Waldhauser and Ellsworth, using a HypoDD program, the earthquake was relocated. The quake triggered a fault bordering the Panama microplate with the Caribbean plate (Pacheco et al., 2006). This earthquake could badly be interpreted as part of the seismic cycle of Costa Rica; however, this was an isolated event. In addition, we believe that the way in which Güendell describes seismic cycles for Costa Rica is incorrect. Seismic cycles are specific to each area of rupture and are affected by changes in the regime of static efforts after each event. For all of Costa Rica, it is not possible to speak of seismic cycles. Large zones of rupture are Osa, Nicoya, and Southern Caribbean, with different return periods of seismic cycles (~40, ~50, and ~70, respectively); therefore, it is possible to have two cycles in Osa without breaking them in Nicoya or the Caribbean, and eventually three seismic cycles could match in a very short period of time.

CHAPTER 2

Central America's Isthmus Geomorphology

From a geomorphological point of view, Central America is characterized by a series of regions formed from a slow but active tectonic and volcanic evolution that occurred during the geological ages. Two major areas there are the Nucleus Central America and the Isthmus Central America. The Nucleus Central America ranges from the Isthmus of Tehuantepec to Northern Nicaragua. It has a long geological history characterized by different orogenic periods (cycles of the Precambrian, Hercynian, and Laramian from the Jurassic to the present distinguished by its great intensity); the current landscape has metamorphic, intrusive, and sedimentary reliefs dating back to the Primary, Secondary, and Tertiary periods. The Isthmus Central America, which is much younger, covers an area from Northern Nicaragua to Panama. It emerged at the end of the Cretaceous period, about 80 million years ago.

It can be said that Central America is the result of the clash of two minor tectonic plates 5 million years ago: Cocos plate that covers part of the Pacific Ocean and the Caribbean plate. This clash created several reliefs through orogenesis and was associated with the magma rising from the mantle through the earth's crust was weakened by the collision of the tectonic plates. This crust weakness allowed the creation of magma chambers, where hot magma passing through fissures created new volcanoes and, in some cases, were completed by eruptions that made existing volcanic structures collapse, leaving vast depressed areas known as *collapsed calderas*. Examples of these event of collapsed calderas are the Managua caldera in Nicaragua, the Ilopango caldera in El Salvador, and the Atitlan caldera in Guatemala. In areas farther from the subduction zone, constant deposits of sedimentation from landslides and general erosion from the emerging mountain ranges resulted in vast flooded plains that gradually created an isthmus between the Pacific Ocean and the Caribbean Sea.

CONTENTS

13

Geomorphology of Central America. http://dx.doi.org/10.1016/B978-0-12-803159-9.00002-9

TEHUANTEPEC ISTHMUS REGION

The southern sector of the Isthmus of Tehuantepec is formed by two Mexican states, Oaxaca and Chiapas, which are in southern Mexico on the Pacific watershed, characterized by very old soils with granodiorite outcrops from the Paleozoic period, is a silent witnesses of a very old relief with partly folded marine sediments of Jurassic and Cretaceous. The sedimentary series continued northward during the Tertiary and Quaternary. The area of the southern sector of the Isthmus of Tehuantepec is crossed by faults oriented northwest-southeast, relatively parallel between them. They created grabens such as the Belisario Dominguez Dam followed by the Grijalva syncline, which is followed by the San Jose anticline and then another syncline, the Chaparral. They resemble a folded area or sinclinorium oriented northwest-southeast and furrowed by a multitude of tectonic faults that indicate a major tectonic activity corresponding to the volcanic arc of Chiapas. There rise 14 Quaternary volcanic edifices, including the cones of the Chichonal or Chichon and the Tacana, the latter bordering Guatemala. The Chiapas volcanic arc is an extension of the Sierra Madre Occidental. The eruption of the Chichonal volcano in 1982 sent ashes over half of the state of Chiapas (Lopez Ramos, 1975) (Figure 2.1).

The **Chichonal volcano** on tertiary sedimentary series dates back to the upper Pleistocene (270,000 years ago) and rises to 3608 ft above sea level in the Sierra de La Magdalena. Its lava was composed of trachyandesites. The crater with a diameter of 1100 yd is occupied by a green-blue pluvial lake. On its borders are fumaroles and sulfur steams. Meanwhile visitors swim on the lake (Figure 2.2).

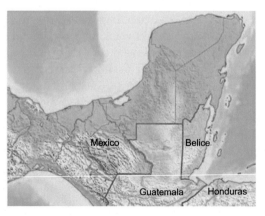

FIGURE 2.1
Isthmus of Tehuantepec in Mexico: The beginning of Central America.

FIGURE 2.2
Crater left by the eruption of Chichonal volcano in 1982, causing 2000 deaths. *Photo courtesy of Universidad de Ciencias y Artes de Chiapas UNICACH.*

YUCATAN AND BELIZE

Central America emerges from North America in the Isthmus of Tehuantepec in Mexico as a new regional geographical unit and extends to the Isthmus of Panama by which Tehuantepec contacts South America, the Yucatan, and Belize, constituting a large geomorphologic unit.

The Yucatan is a large sedimentary plateau formed by sedimentary rocks of the Mesozoic (limestone, sandstone, and evaporite; Lopez Ramos, 1975) with a thickness of 11,500 ft resting on a Paleozoic basement. The outcropping rocks belong to the Cretaceous. The Yucatan is characterized by two main morphological units: a northeastern portion with an altitude of 164 ft and a southern portion with an altitude of 1300 ft. The area of its lithologic composition was affected by the karst dissolution phenomenon with sinkholes, or cenotes, such as Sac-Actun or Dos Ojos. The term *cenote* from the Maya ***dzoonot*** refers to a deep hole in the ground that can reach 300 ft or more that is flooded by an aquifer. It is the typical morphology form of the northeast of Yucatan (Figure 2.3).

In the north of the Yucatan Peninsula is an enormous crater created by the impact of the Chicxulub meteor, which was 6 mi in diameter. Discovered by geophysicists Antonio Camargo and Glen Penfield at the end of the 1970s, the crater is 112 mi in diameter. Analysis of the rocks showed that they contain

FIGURE 2.3
Cenote of Dos Ojos, north of Tulum, tourist dive center. *Photo courtesy of Ulises Grajales Valdivia, 2012.*

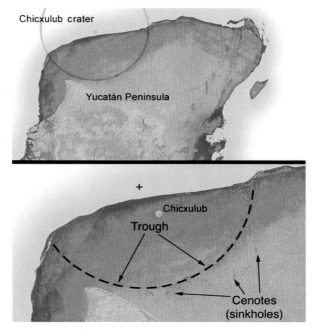

FIGURE 2.4
Northern Yucatan Peninsula where Chicxulub meteor impacted the earth 65 million years ago. *Image courtesy of http://en.wikipedia.org/wiki/Chicxulub_crater#mediaviewer/File:Yucatan_chix_crater.jpg.*

iridium, a rare element on earth, as well tektites; isotopic studies indicated their age to be 65 million years at the boundary of Cretaceous and Tertiary periods. This impact and that of others that accompanied it are thought to have been responsible for the mass extinction of saurians during these periods (Figure 2.4).

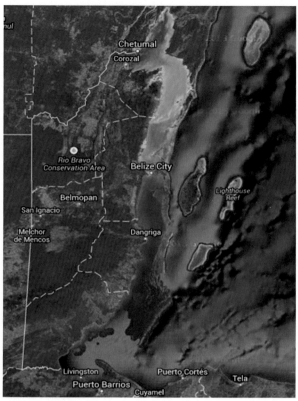

FIGURE 2.5
Satellite map of Belize on the Caribbean Sea between Mexico in the north and Guatemala to the west and south. *Satellite Google map image, 2014.*

Belize, now an independent country in Central America, covers square 8878.8 mi^2 and is bordered by Mexico, Guatemala, and the Caribbean Sea. Its coast, covering 228 mi, is formed by coral reefs and a key line forming the Turneffe Islands. Yucatan and Belize formed part of the Mayan Empire until the arrival of Christopher Columbus in 1502 and became the Viceroyalty of New Spain. In 1630, English buccaneers took over and in 1713, Spain recognized it as a British colony, which explains why it is the only English-speaking country in Central America. It exported logwood (palo campeche) (Figure 2.5).

Belize is composed of several geomorphological sets. In the south and center of the country, a very eroded Primary relief protrudes from the humid low plains. This granite relief constitutes a massif that includes the Maya Mountains, which reach a maximum altitude of 3674 ft in Victoria

FIGURE 2.6

The Great Blue Hole, located 60 mi off Belize coast on the Lighthouse Reef, is a sinkhole of karstic origin that must have occurred during the last ice age (Würm-Wisconsinan) when the sea level was 460 ft below the current level. It is now 663 ft deep and 984 ft wide. Karstic stalactites appear in its depths. *Image courtesy of http://www.oceansub.es/la-gran-barrera-de-coral-hd/gran-agujero-azul-belice-2/.*

Peak. The southern low coast is covered by mangroves southward from Chetumal Bay to Amatique Bay on the southern border with Guatemala. Northern Belize is characterized by low hills formed of Tertiary limestone and karst dolines. The northern coast has small sandy bays and coral cays (Figure 2.6).

CHAPTER 3

Geomorphological Landscapes of Guatemala

Guatemala is a country of great physical contrasts, where the Peten jungle plains are juxtaposed with the vigorous mountain relief of the Alto Cuchumatanes Range and the Quaternary volcanic range, descending steeply on the Pacific coastal plain.

THE HUMID PETEN TROPICAL PLAIN

The great Peten plain is located in northeast Guatemala and covers an area of 13,843 mi^2, covered with a thick rainforest that rests on floods deposited from the Cretaceous and Tertiary periods to the present day. The floods are brought by the large hydrographic network that runs through the plain, particularly the Usumacinta River, which forms the plain's west Mexican border. The subsoil is characterized by a sedimentary series, with abundant limestone of the Cretaceous period that gives it particular karst morphology, shown by collapsed doline landscapes called *"Cenotes"* by the Mayan. The area emerged from the sea during the Miocene, about 15 million years ago. Four large geomorphological sectors dominate this region, based on its lithology and modeling.

Yucatan vast platform. This area extends from the north to the vicinity of Peten Itza Lake. It is a low plateau, formed by monocline series of sandstones and limestone, marls and plaster, belonging to the period from the late Cretaceous to the Eocene. It is an extension of the Yucatan Peninsula and covers fine sediments in rivers and decomposing material in situ. It is subject to deep karstic dissolution.

Peten Itza Lake central depression. The subsoil is composed by the Lacandon folded Cretaceous sedimentary series that conditions the east-west elongated Peten Itza Lake, which was formed to the east in contact with the first foothills of the Mayan Mountains. The lake is a natural intermediate boundary between the Yucatan platform occupied by agricultural work and the south dominated by the karstic relief covered with tropical vegetation. Situated 109.36 yd above sea level, the lake covers an area of 38.224 mi^2; it is the third largest lake of Guatemala, after Izabal Lake and Atitlan Lake. Its maximum depth reaches

CONTENTS

19

Geomorphology of Central America. http://dx.doi.org/10.1016/B978-0-12-803159-9.00003-0

FIGURE 3.1
Peten Itza Lake and Noh Peten City, in Flores Island. *Aerial photography courtesy of www.Oirsa.org.*

184.8 yd. Currently, Peten Itza Lake levels rise gradually because the lake receives more water, which evaporates. It is home of the Itzaes Mayas, who named the lake "waters of the witcher." They were the last aborigines to be conquered by the Spaniards (Figure 3.1).

Southern dissolution karst plateau. This plateau extends to the south of Peten Itza Lake. Same high reliefs, though profoundly altered, is constituted by a series of limestone dissolution under karst. The results are a combination of tropical rainfall, associated with the acidity introduced by the decomposition of the tropical vegetation in a basement of calcium magnesium carbonates, dating back to the late Miocene. This plateau, formed in the sea, emerged during the latest orogenesis.

Mayan mountain range. Located in northeastern Peten, the Mayan mountain range is the highest area where few agricultural activities are implanted, because most of the territory is covered by tropical rainforest. The altitudes vary from 656.17 to 1312.3 yd. It is a landscape of rolling mountains, forming a tropical, multi-convex relief. Its origins date back to the Carboniferous and Permian periods (Figure 3.2).

IZABAL LAKE TECTONIC SYSTEM

Izabal Lake is a tectonic deformation sector due to the presence of the transforming **Motagua-Polochic fault**. It has an arched movement of left lateral direction and is inserted into the fault system, which marks the boundary between the North American plate (to the north) and the Caribbean plate (to the south). The rocks that make up this unit date back to the Permian (Cholchal formation). The sector is consistent with the orogenic deformations

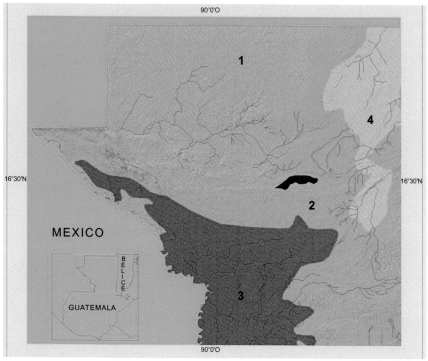

FIGURE 3.2

Peten and Peten Itza Lake sectors: 1—Yucatan karstic plateaus (Cretaceous); 2—Folded Lacandon formations (Cretaceous); 3—Mayan interior plain (Miocene-Tertiary); 4—Mayan Mountain reliefs (Carboniferous to Permian). *Photo interpretation by J.P. Bergoeing, 2011, based on Unipesca, Guatemala, 2010.*

of the Paleozoic geosynclines (Carboniferous and Permian), dominated by serpentine and meta-volcanic rocks that were later folded and faulted.

Left diverging failures of Motagua and Polochic are complex. They would have 9-mi depth and a displacement speed of $0\,{}^{25}/_{32}$ in/year, and would be reduced to $0\,{}^{19}/_{32}$ in/year in the sector of the Triple Union (Caribbean, Cocos, and North American plates) (Franco, 2008). The Motagua-Polochic fault is the tectonic boundary between the North American plate, as previously mentioned, and the Caribbean plate forms the **Southern Chortis block**. In this sector, the fluviolacustrine system of Izabal Lake focuses on its tributaries—the Polochic River (149 mi. long), Dulce River, and the Golfete—before flowing into the Amatique Bay (Caribbean Sea), where the main Caribbean port of Guatemala, Puerto Barrios, stands. Izabal Lake, which is $6691\,{}^9/_{64}$ in. deep (Brooks, 1969), is inserted in a tectonic depression bounded on the north by the Motagua failure and the Santa Cruz Mountains, and on the south by Las Minas Mountains. It is therefore a natural collector of the rainwater channeled by the numerous tributaries flowing into the Izabal Lake. The description of this sector would be incomplete without mentioning Motagua River, which after a journey of 248 mi, flows into the Honduras Gulf and remains the longest in the country (Figure 3.3).

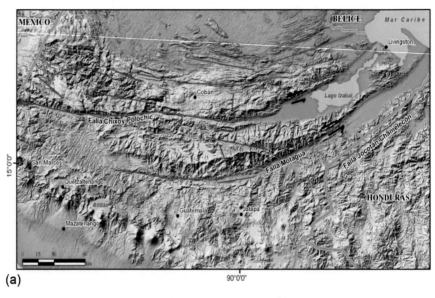

(a)

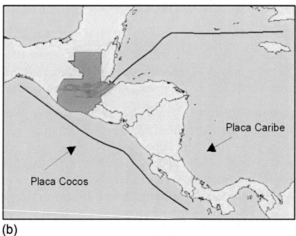

(b)

FIGURE 3.3

Motagua, Xichoy-Polochic, and Jocotan-Chamelecon transforming faults in Guatemala. Tectonic limits between North American, Caribbean, and Cocos plates. *Photo interpretation by J.P. Bergoeing, 2012. Source: Geological map of the Guatemalan Republic (MAGA-IGN, 2001).*

ALTO CUCHUMATANES, A GLACIAL PAST

The Huehuetenango Department in Guatemala stands out for the Cuchumatanes Mountains' end of the Sierra Madre Occidental Mountains of Mexico, which separates two slopes: a southwestern slope, where strong rivers descend, and a northeastern slope, birthplace of the Usumacinta and Selegua Rivers, heading to Mexico. It is also characterized by the highest summits in the country; among them is Alto Cuchumatanes Mountain, which exceeds 11,482 ft. This gives it the characteristics of having suffered the last Quaternary glacial layer. Deposits of glacial moraines, erratic blocks, and layers of ice were studied by several researchers (Anderson, 1969; Hastenrath, 1973; Lachniet, 2011). The latter determined that Wisconsinian parking ice covered a limestone plateau, north of Todos Los Santos de Cuchumatanes, covering about 23.166 mi^2 (Lachniet and Roy, 2011; Lachniet and Seltzer, 2002). The limestone plateau of Los Cuchumatanes is characterized by a karst undulating relief with dolines (sinkholes), which is dominated by an underground drainage. Some valleys possess evidence of glaciations with moraine deposits. Miles from San Sebastian of Huehuetenango and Todos Los Santos, there is evidence of a terminal moraine and a tarn of a former glacial lake. At the end of the Pleistocene, as in Talamanca Mountain in Costa Rica, the Cuchumatanes region was elevated enough to find the necessary climatic condition for the development of glacier deposits (Lovell, 2005) (Figure 3.4).

GUATEMALA'S SIERRA MADRE, VOLCANIC RANGE

Of the 324 volcanoes indentified in Guatemala, only 8 of them have had eruptions in historical times, and 4 of them have had great activity. Most of them are stratovolcanoes. The first volcanic cone rises in the northern border with Mexico. It is the **Tacana** or **Soconusco** volcano, located in San Marcos Department, reaching an altitude of 13,425 ft—one of the highest Guatemala volcanoes. It has had eruptions in the last century; the last one occurred in

FIGURE 3.4

Alto Cuchumatanes Mountain paleoglacier modeling. Ninguitz de La Ventosa Valley. Lateral and terminal moraines. Small tarns, center of image. *Photo courtesy of Alex Roy, 2010.*

1986. It is an active stratovolcano, emitting andesitic lava flows. It began to form in the Middle Tertiary on a granite basement. It presents fumaroles at its southwest flank, as well as springs. It has had phreatic eruptions and periodic fumaroles registering since the nineteenth century. Close to the Tacana emerges the imposing **Tajamulco** stratovolcano, which rises to 13,845 ft, but verified historical eruptions are not known. It has an open crater to the west and the rest of an explosion caldera (Figures 3.5 and 3.6).

Santa Maria or Gaxanul Volcanic Complex

Consisting of a series of volcanic cones, **Santa Maria** is the most important volcano, both by its activity and its elevation (12,375 ft). It is the most active volcano in Guatemala; its Plinian eruptions are exceptionally important and explosive. The northern flank is dominated by two stratovolcanoes. **Almolonga** (10,488 ft) is situated in the northeast, with an open crater to the west and an important recent lava flow to the east, where a powerful lava flow at the Almolonga city probably emerged. **Siete Orejas** volcano (11,056 ft), which stands to the northwest, was built on a former explosion caldera that is highly eroded today. East of the Siete Orejas emerges **Chicabal Lake**, which is a volcanic crater full of pluvial waters (Figure 3.7).

To the east of this complex emerges the **Santo Tomas** or **Pecul** stratovolcano of 11,620 ft. It is the oldest volcano of this set, and therefore was the first to be created during the Upper Tertiary. It covers a large area to the west of these volcanoes. Santa Maria had four catastrophic eruptions during the twentieth century (1902, 1912, 1992, and 1994). The 1992 eruption destroyed part of

FIGURE 3.5

Tajamulco volcano and fires of human origin in its summit. *Photo courtesy of Arnoldo Marroquin, Prensa Libre, Guatemala, 2013.*

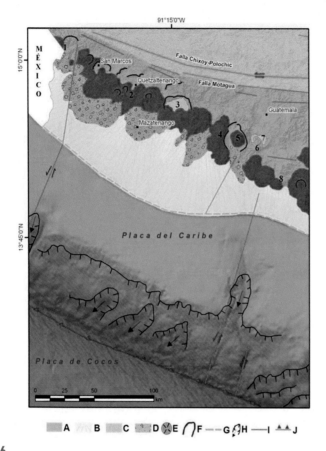

FIGURE 3.6

Guatemala "Sierra Madre" Quaternary volcanic alignment. A—Sierra Madre's volcanic range (Tertiary-Quaternary); B—Alluvial sediments (Quaternary); C—Undifferentiated Secondary-Tertiary area; D—Lahars; E—Volcanic focus: 1—Tacana, 2—Santa Maria, 3—Atitlan caldera, 4—Fuego volcano, 5—Agua volcano, 6—Pacaya volcano, 7—Amalitlan caldera, 8—Tecuamburro volcano; F—Calderas; G—Littoral strips; H—Under seawater landslides; I—Tectonic faults; J—Cocos plate subduction area. *Geomorphological photo interpretation by J.P. Bergoeing, 2012.*

its crater, leaving a large scarp and allowing the emergence of the **Santiaguito** cone (8202 ft), where a constant plume of ash and steam escape from its crater, rising more than 2 mi above sea level. The Santa Maria volcanic complex is part of the Southern Sierra Madre and is a result of the Central American volcanic arc. It is estimated that its activity began about 30,000 years ago, which is very recent (Upper Pleistocene). The lava flows are predominantly basaltic-andesitic, and also present numerous lahar deposits as a result of pyroclastic and cinder emissions associated with tropical rains. At the foot of this complex stretches the city of Quezaltenango.

FIGURE 3.7
Santiaguito's dacitic dome in eruption, forming part of the Santa Maria volcanic complex in the background. *Photo courtesy of Jessica Ball, Department of Geology, University of New York.*

Atitlan Caldera

Atitlan caldera is located in the Solola Department. It is actually a semicircle occupied by a lake that covers an area of 48.649 mi². In the southern sector emerge three stratovolcanoes. In the southwest, **San Pedro** volcano, which is 9908 ft above sea level, is separated from **Toliman** volcano (10,360 ft) by an entry of the Atitlan Lake where the cones of both volcanoes have not yet coalesced. South of the Toliman emerges the **Atitlan** volcano (11,604 ft), with known eruptions from 1469 to 1856. This set of three volcanoes is the reconstruction in course of the old collapsed volcanic caldera complex of Atitlan. If these volcanoes are Quaternary, the unit is seated on a previous Pliocene volcanism (Figure 3.8).

Fuego Volcanic Mountain Range

Fuego volcanic mountain range is formed by five stratovolcanoes reaching an average altitude of 11,482 ft. It is composed by the cones of the **Atenango** (12,730 ft), **Yepocapa**, **Pico Mayor**, **Meseta**, and **Fuego** volcanoes.

Fuego Volcano

This stratovolcano rises to 12,345 ft, forming a perfect adjacent cone with the **Acatenango** cone, and dominates Antigua City. A vast plateau developed to the south with avalanches' slag deposits. It has violent, Vulcanian-type eruptions, with a historical record since 1524. Since then it has had more than 60 eruptions, most recently in 2012. The volcanic focus of Fuego is simply the result of magma migration of the Acatenango, whose lavas are basaltic. Near this volcanic complex rises **Agua volcano** (Figure 3.9).

FIGURE 3.8
Atitlan collapsed caldera and Lake San Pedro volcanic cone at right; Toliman volcano at left in the foreground; Atitlan volcano behind. All the volcanoes are in activity. *Photo courtesy of Tom Pfiffer, www. volcanodiscovery.com.*

FIGURE 3.9
Fuego volcano's pyroclastic flows descending to the base of the cone during the September 13, 2012 eruption. *Photo courtesy of Elmundo.com.sv.*

Agua Volcano

The Agua stratovolcano rises to an altitude of 12,335 ft, east of the Fuego volcanic complex. It has been inactive since the sixteenth century. It has a northward opening crater, forming a perfect cone that is bordered by a much-eroded old volcanic structure, which is located between the cities of San Pedro de Las Huertas and Amatitlan, and which corresponds to an old volcanic cone.

Pacaya Volcano

Pacaya volcano reaches 8372 ft and its last eruption dates back to 2000, with an average of 50 years between eruptions. The oldest eruption dates back to 1565. The modern top was formed on an explosion caldera that is home to a pluvial lake near the town of Bejucal. Two modern cones were formed inside the caldera. The northern cone has an inactive crater covered with vegetation. The southern active crater emits basaltic lava flows toward the south and directly threatens the Los Pocitos village, where most of the recent lava flows have been directed. To the north, the volcano borders Amatitlan Lake. Farther north, Pacaya volcano is limited by a second collapsed caldera, where the town of San Vicente Pacaya is located and which houses a series of craters, of which **El Cerrito** Crater Lake, close to Amatitlan Lake, is the best preserved. If the main cone is covered by recent lava flows with a dark color at the top, the lava flows are directed mainly toward the south (Los Pocitos sector). This is according to the structure of the cone being limited northward by the described caldera.

Amatitlan Lake

Amatitlan Lake occupies a collapsed caldera depression 15 mi south of Guatemala City, dating back about 40,000 years (Upper Pleistocene). It is situated 3891 ft above sea level, covering an area of 7 mi 803.36 yd long and 1 mi 1520.8 yd wide. The caldera is occupied by a lake whose waters cover approximately 5.8688 mi^2, with a maximum depth of 36.089 yd. Amatitlan caldera belongs to the Pacaya, Fuego, Agua, and Acatenango volcanic complex, covering a total area of 30.888 mi^2. The lake is fed by many tributaries, with the Villalobos River having the highest flow. The lake is drained by the Michatoya River tributary of the Maria Linda River, which flows into the Pacific Ocean. Much of the lake is occupied by a large alluvial fan that forms a lake delta, which starts in Villa Canales, and where the progression of sedimentation threatens to divide the lake into two parts. In fact, the northeast part of the lake is an important lacustrine delta formed by sedimentary input from tributaries that feed the lake, which is responsible for the narrowness of the lake. It will probably end up dividing and creating two separate lakes unless humans intervene. The delta is quite visible in Figures 3.10 and 3.11.

Sinkholes in Guatemala City

Recently, Guatemala City suffered the collapse of ancient sinkholes covered by street tar, leaving gaping holes. This was the result of filling stone pumice and ash deposited by eruptions of the neighboring volcanoes in an ancient valley, where the city was later built. This part of the city was built on a former pumice stone soil. Pseudokarstic sink holes in Guatemala City formed from 2007 to 2010 as a result of water infiltrations in a basement composed of pumice and volcanic ash accumulated during eruptions of the nearby volcanoes. These holes can reach 200-400 ft deep and 100 ft wide. They can occur anywhere and anytime in the city. They have a similar morphological behavior to karst but, in this case, in a volcanic environment (Figure 3.12).

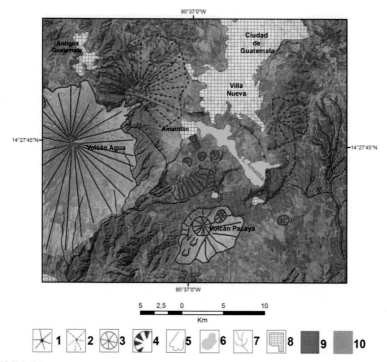

FIGURE 3.10

Amatitlan Lake area, volcanic geomorphology. 1—Volcanic cones; 2—Old eroded volcanic cones; 3—Volcanic domes; 4—Collapsed caldera rims; 5—Recent lava flows; 6—Amatitlan Lake area; 7—Hydrological frame; 8—Urban areas; 9—Upper Pleistocene volcanism; 10—Holocene volcanism. *Photo interpretation by J.P. Bergoeing, 2012.*

FIGURE 3.11

Lacustrine delta and Amatitlan caldera, as seen from the southeast. In the background, the Pacaya volcanic cone drops steeply into the lake. *Photo courtesy of Jim Reynolds, 2009.*

FIGURE 3.12
Ten sinkholes 100 ft deep appeared from April 2007 to June 2010 in Guatemala City. *Photos courtesy of Luis Echeverría, 2010. EFE/Secretaria de comunicacion de la Presidencia.*

Tecuamburro Volcano

Thirty-one miles southeast of Guatemala City emerges the **Tecuamburro or Miraflores** stratovolcano, with an altitude of 6053 ft. It is an andesitic volcano that formed at the end of the Pleistocene, about 38,000 years ago, from a collapsed caldera dating back 100,000 years. At its summit crater, the Tecuamburro is home to an acid lake with fumaroles and mud emissions (Figure 3.13).

Collapsed Calderas
Ayarza Collapsed Caldera

The Ayarza collapsed caldera represents remnants of volcanic structures between El Aguacate and Tatasire, and is located in eastern Guatemala City. However, it is the **Ayarza Lake** that impresses most, as it is a collapsed caldera that blew up about 20,000 years ago, leaving large amounts of pumice deposits. The

FIGURE 3.13
Tecuamburro stratovolcano with an altitude of 1 mi 257.72 yd. *From Wikipedia.*

edges of the lake are steep and draw a double caldera, which would indicate the former existence of two collapsed volcanic edifices. Following the geology of the area, the northern sector is older and goes back to the Pleistocene, while the southwest sector is much more recent, dating from the Upper Pleistocene (ISRIC World Soil Database) (Figure 3.14).

Retana Collapsed Caldera

To the northeast of Juliapa City, among the **Tahual** (5629 ft) and **Suchitan** (6699 ft) stratovolcanoes, extends the vast **Retana collapsed caldera**, which is currently covered by crops. The caldera was formed by basaltic-dacitic lava surrounding the central circular depression. North of the caldera, following

FIGURE 3.14
Ayarza Lake, in a double collapsed caldera from 20,000 years ago. Vertical walls are formed by pyroclastic and dacitic materials. *Photo courtesy of Ivan Castro, Guatemala, 2007.*

a north-south alignment, extends a series of dacitic domes and gas maars (phreatic-magmatic explosion crater lakes) (Figure 3.15).

The Southern Volcanic Structures
Tahual Volcano

Tahual volcano rises 5629 ft and is characterized by an open crater to the northeast, where it has fitted a deep river canyon using the northeast-southwest tectonic fault alignment, and at its feet drawn a series of gas maars as **Laguna del Hoyo Lake**. Also, the **Suchitan** stratovolcano (6699 ft) is present as a much-eroded cone, which nevertheless possesses the **Metaltpeque adventitious cone** (6082 ft) located in its upper flank, and represents a second eruptive phase after the activity cessation of the first crater. It is characterized by two equally important protohistorical basaltic lava flows and pyroclastic deposits.

Flores or Amayo Volcano

Flores stratovolcano reaches an altitude of 5249 ft and is located 6 mi from Jutuiapa City, dominating the urban centers of Santa Gertrudis and Quezada. It was formed on a Cretaceous-Tertiary sedimentary basement and is characterized by a series of adventitious cones that emerge in their southern base, as well as by recent basaltic lava flows. The main crater at its summit is a cone whose crater is plugged and covered with vegetation.

Moyuta Volcano

Moyuta stratovolcano, rising to an altitude of 5452 ft, marks the last volcanic structure of Guatemala before entering the volcanic arc of El Salvador. Historical volcanic activity is not known for this volcano. It lies to the south of the

FIGURE 3.15
Retana collapsed caldera. The bottom of the caldera is covered by crops. The surroundings present the circular rim and a volcanic cone in the background. *Photo courtesy of GalasdeGuatemala.com.*

FIGURE 3.16
Moyuta stratovolcano in Santa Rosa Department, situated at the southern edge of the Jaltapagua fault. Its summit is formed by three andesitic lava domes. *Wikipedia.*

Jaltapagua fault, dominating the graben of the same name. It has numerous cinder cones and andesitic lava domes. On the base of its northeast and southeast flanks are the nearby hot springs of Azulco and Ausoles. Also, its northern border sector with El Salvador is characterized by another complex of three stratovolcanoes. The **Ipala volcano** (5413 ft) has a diameter crater of 0.5 mi with vertical walls of 492 ft, and hosts a pluvial lake. It forms part of the cineritic cone field of Southern Guatemala. This volcano is followed by the **Suchitan** and **Tahual** volcanoes, which have already been described. These three volcanoes, bordering the northwest of El Salvador, mark the transition with the **Metapan** complex dominating **Guija Lake**, which is another collapsed caldera (Figure 3.16).

THE PACIFIC PLAIN

Parallel to Guatemala's volcanic system extends the Guatemala Pacific Coast plain, about 250 mi between the borders with Mexico from the north to the south with El Salvador, following a northwest direction. It is a narrow strip that does not exceed 37 mi wide, and corresponds to the volcanic foothills of the Quaternary system of the country. The coastline is a lobular or convex linear coast toward the Pacific Ocean, without natural bays for port construction, and is therefore sparsely populated.

This product of volcanic landslides and dense mangroves is characterized by black sand beaches. In the interior plain are sugar cane, cotton, and cocoa crops. The most important sector is the **Puerto Quetzal** and **Itzapa** ports on the banks of the Chiquimula canal, where a vast mangrove swamp stretches, corresponding to the Monterrico flora and fauna of the Natural Preservation

Reserve. The vast continental shelf does not exceed 124 mi from the coastline before reaching the continental slopes that in some sectors can reach 15,091-16,076 ft deep, which is a result of the subduction of the Cocos plate colliding with the Caribbean plate. It is in front of the **Puerto San Jose** port artificially built in the floodplain, 3 mi west of Puerto Quetzal, that we find an abyssal pit perpendicular to the Guatemalan Pacific abyssal pit. This is a product of tectonics associated with the deep sea currents of the sector. From this point northward, we also find the volcanic complex system of Fuego, Agua, and Pacaya.

The Piedmont, which ends in the Pacific coastal plain, is extremely regular as it moves northwest. If you take the Santa Maria volcanic complex to the Pacific coastal area (Champerico), it is possible to observe a multitude of rivers descending from this complex through huge coalescent alluvial fans that flow into the Pacific Ocean through populated mangrove ponds. Here we can see La Maquina agricultural development zone adding value to these agricultural lands. Farther southeast, between the towns of Madre Vieja and Buena Vista, stands the Monterrico-Hawaii biotope zone, which has the vastest mangroves in the sector (about 15 mi long). In Barra de Santiago, another vast mangrove area covers 8 mi as a continuation of the first one. These vast mangroves are the joint result of littoral drift currents that have guided the mouths of many rivers reaching the Pacific Ocean, forcing them to take a parallel course to the coastline, thus allowing the proliferation of riparian vegetation before becoming a mangrove. The straight line of the Guatemalan Pacific Coast stops in El Salvador, bordered by the Acajutal River delta (Figures 3.17 and 3.18).

FIGURE 3.17
Beach formed by a littoral cord in the Pacific Coast that protects the adjoining mangrove. Escuintla Department, Guatemala. *Photo courtesy of Cb24tv.2014.*

FIGURE 3.18
Quaternary volcanic range merging in the Guatemalan plateau: Agua volcano, Pacaya volcano, and then Fuego volcano. *Photo courtesy of Panoramio Google maps.*

CHAPTER 4

Geomorphological Features of Honduras

Honduras is a vast territory that occupies an area of 43,433 mi^2. It belongs to the Chortis tectonic block, separated from the Mayan block by the Motagua-Polochic-Jocotan fault system. Its geomorphologic features allow us to divide the country into several large sections (Figure 4.1).

CARIBBEAN SEA CORAL ARCHIPELAGO

The archipelago is formed by the **Bay Islands (Utila, Roatan, and Guanaja)**, as well the island of Swan and Cochinos keys of Bobel, Port Royal, Savannah, and South. It is a set of small, well-preserved coral islands, and is therefore a preferred tourism destination and fishing zone (Figure 4.2).

However, in this sector of coral islands, the **Utila coral-volcanic island** emerges. East of the island complex, two pyroclastic cones rise in Utila: Stuart Hill (in the center of Utila Town) and Pumpkin Hill (reaching only 242 ft). They are volcanic deposits of the end of the Pleistocene, composed of andesitic lava flows resting on coral reefs as well as tuffs (Smithsonian Institution, GVP, volcano information). The interior of the island presents a myriad of swamps and dolines. The vegetation is characterized by mangroves, with outstanding species like *Laguncularia racemosa*, *Veranus indicus*, *Emeia trocostata*, and *Boija dendrofila*, to name a few (Gutsche, 2005). The coast is bordered by strips of coral reefs, and in the south, a barrier coral reef formation emerges (Figure 4.3).

THE NARROW PLAINS

The narrow plains are located in northern Honduras bordering the Caribbean Sea, and are progressively widening eastward. The extend 670 mi from the Guatemalan to Nicaraguan borders. Here stand the Caribbean ports of Puerto Cortes, La Ceiba, and Trujillo, where the country's main export ports are. Alluvial plains are fed by many rivers that flow into the Caribbean, including Chamelecon, Ulua, Aguan, Sico Tinto Negro, Patuca, and Segovia (or Coco) (Figure 4.4).

CONTENTS

37

Geomorphology of Central America. http://dx.doi.org/10.1016/B978-0-12-803159-9.00004-2

FIGURE 4.1
Fantasy Island Resort in Roatan, Honduras. System of strips of coral keys in the Caribbean, which have allowed the formation of a beach and interior lagoons belonging to the coral keys system. *Photo courtesy of Mark Steven Houser, 2010.*

FIGURE 4.2
Utila Island, Arenoso Bay, and Utila Town surrounded by ponds and strips of coral reefs. The pyroclastic Pumpkin Hill volcanic cone is on the bottom. *Photo courtesy of www.Parrotsdivecenter.com.*

Ulua River

Among the Honduran rivers flowing into the Caribbean Sea, one of the most important is the Ulua River. It has a watershed of 8.4803 mi² and crosses 10 departments, watering the Sula Valley. Born in the Intibuca Department under the name Grande de Otoro River, it runs about 248 mi until the sea, and it can vary from 21 to 62 mi wide. Its main tributaries include Jicatuyo, Otoro, Humaya, Sulaco, and White Rivers. In its middle course, it receives equally numerous tributaries, increasing its flow and usually causing flooding,

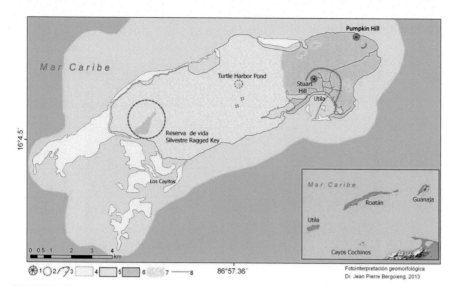

FIGURE 4.3

Utila Island geomorphology: 1—Pyroclastic cones; 2—Dolines or sinkholes; 3—Volcanic eruption area; 4—Strips of coral reefs and barrier coral reefs; 5—Swamp area; 6—Volcanic deposits; 7—Mangroves; 8—Urban area. *Photointerpretation by J.P. Bergoeing, 2013.*

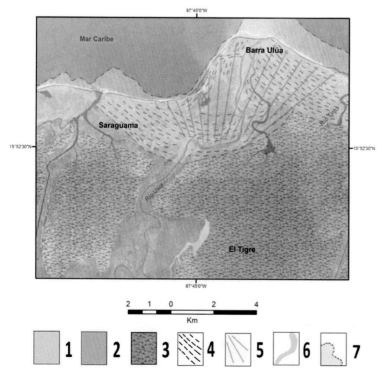

FIGURE 4.4

Ulua River lower watershed geomorphology, Honduras: 1—Fluviomarine sediments (Holocene); 2—River sediments formed partly by the Ulua River (Middle to Upper Pleistocene); 3—Coastal dense tropical forest; 4—Flandrian costal ridges; 5—Alluvial fan passing to a delta formed by the Ulua River and others draining into the Caribbean Sea; 6—River network; 7—Sea sedimentation area composed by colloids in suspension, contributed mainly by the Ulua River and displaced by a littoral drift current. *Photointerpretation by J.P. Bergoeing, 2013.*

FIGURE 4.5
Ulua River, Sula Valley, Honduras, describing meanders in its low course. *Aerial photo courtesy of http:// www.xeologosdelmundu.org.*

particularly during hurricane season. In its margins, prosperous farming and coffee crops are developed. At its river mouth in the Caribbean Sea, Ulua River forms a sandy bar, despite having a single arm that describes numerous meanders in its low-course divagation, forming a delta crisscrossed by numerous Flandrian coastal sandy bars that have left gaps inside, a witness to recent fluviomarine sedimentation (Figure 4.5).

THE GREAT CARATASCA COASTAL ALLUVIAL PLAIN

Caratasca plain lies to the northeast, bordering Nicaragua and corresponding to the **Mosquitia** territory. It stands out as a flooded region, marshy and covered with lakes. The coast is surrounded by lagoons carved in fringing coral reefs. This sector is crossed by the 62-mi **Coco River**, which marks the border with Nicaragua and covers an area of 8.4803 mi^2. Coco River is the most important in this sector. Its river mouth is a vast delta ending in a point just into the Caribbean Sea, and constitutes **Cape Gracias a Dios**. The river mouth has been constructed by numerous Flandrian coastal ridges, a product of the littoral drift encountering currents, letting a large number of coastal lakes inside. Some of these lakes have large dimensions, such as the Caratasca Lagoon, where Puerto Lempira port stands. It is possible to access the Caratasca Lagoon from the Caribbean Sea by a narrow bar (Figure 4.6).

This lake is fed by numerous tributaries, including the Warunta River, flowing like many others on the remains of a much-eroded Jurassic and Cretaceous

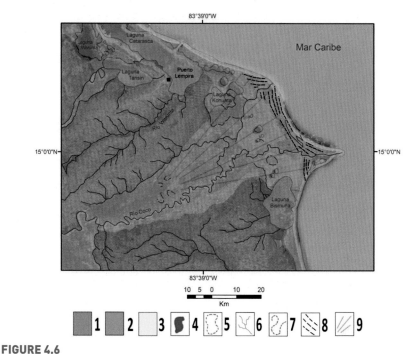

FIGURE 4.6

Caratasca plain geomorphology. Coco River low course, international border with Nicaragua: 1 and 2—Cretaceous to Tertiary worn reliefs; 3—Quaternary sediments; 4—Coastal lakes; 5—Disappearing lakes; 6—Hydrological network; 7—Meanders; 8—Coastal cords area (Holocene); 9—Coco River alluvial fan. *Satellite photo interpretation by J.P. Bergoeing, 2013.*

hill relief. The sector was described by 19th century traveler **Ephraim George Squier**, whose "Appointments of Central America" was published in Paris in 1856 by editor Gustave Gratiot. This is what he says in reference to the Caratasca Lagoon, which in the Miskito language means "Lizards Lake" (Figure 4.7):

> The lagoon of Caratasca or Carthage is of considerable extent, varying wide and taking in some places the appearance of several parallel coast lagoons gathered in different directions, but not extending of twelve miles wide. It has two entrances; one of them is a small Cove called Tibacunta. The main mouth is wide, with thirteen or fourteen feet of water at the bar. The lagoon is estimated at thirty-six miles of extension. For the most part, it is dry, varying in depth from six to twelve or eighteen feet. Captain Henderson, who visited it, describes the country immediately to the Sambo village or Carthage "like a spacious sheet", forming a full level covered of vegetation and good pasture, cut on one hand by the waters of the lagoon and on the other by high hills...

FIGURE 4.7
Caratasca Lagoon extends 25 mi inland from the Caribbean Sea, and measures more than 55 mi from northwest to southeast. It is linked to the Caribbean by a 3-mi channel, on the bank of which stands the Caractasca village. *Google Earth satellite image, 2014.*

THE CENTRAL RANGE

The central mountain range occupies most of the Honduran territory. It is composed mostly of volcanic, sedimentary, and metamorphic formations of the Primary and Secondary eras, as well as some intrusive rocks of similar ages that alternate with Jurassic and Cretaceous limestone. Farther to the south, mountain ranges are composed of younger volcanic rocks of the Tertiary. The central mountain range can be subdivided into four large mountain areas: three that have a general east-west direction, and an older one that is oriented north-south and is located on the eastern side of the country.

The North Mountain Range

The north mountain range contains the oldest mountains of Honduras, formed by metamorphic outcrops of gneiss and slates, associated to the granitic batholitic shield. Volcanic rocks are where basalts and Paleozoic and Mesozoic sedimentary rocks prevail, all of which are subject to intense folding of the Earth's crust and tectonic faulting parallel to the path of the Chamelecon, Ulua, and Aguan Rivers. Outliers are part of the nucleus Central America and are disposed in three principal ranges:

>   **Sierra del Espiritu Santo**
>   **Sierra de Omoa**
>   **Sierra de Nombre de Dios**

Here rises the highest summit of this region, Pico Bonito (7988 ft).

The Central Mountain Range

The central mountain range constituted the intracontinental depression during the Cretaceous. It is made up of sedimentary rocks reaching considerable thickness—up to 19,685 ft. It is folded and cut by numerous tectonic faults, a product of the **Laramian Orogeny**. Limestone is abundant, which was affected in part by karst processes (Atima Formation). The central mountain range is crossed by many ranges, including the **Sierra Gallinero**; the prestigious **Copan Mayan ruins** are located in the Santa Rosa plateau. **Yojoa Lake** is located in Sierra de Monticello, between the Departments of Comayagua, Cortes, and Santa Barbara. The lake rises at 395 mi, surrounded by peaks higher than 8530 ft above sea level, and covers an area of 34.363 mi^2 and 98-ft depth. Its origin is a volcanic obstruction lake of the end of the Tertiary, surrounded by folded Cretaceous limestone (Yojoa Group) and acid volcanic rocks of the same age.

The South Mountain Range

Formed by the **Celaque Mountains**, the south mountain range is composed of metamorphic rocks, covered by Jurassic-Cretaceous sedimentary series, folded during the Tertiary and volcanic inputs of the Tertiary. During this period, the south mountain range was the southern limit of the intracontinental depression. Equally constituted by numerous ranges, the south mountain range is characterized, among others, by the 249-mi **Sierra de Dipilto**, which is the most extensive of the country. It covers the Choluteca Department, heading for the border with Nicaragua, which covers most of the sector.

Patuca Mountain Sector

The Patuca mountain sector is oriented north-northeast-south-southwest and is mainly composed of Paleozoic rocks (Cacaguapa schists) corresponding to the older sector of Honduras. It is in the Patuca sector, 3 mi from Catacamas village, where glittering skulls and skeletons of **Pech** indigenous culture were discovered in the Talgua caves, dating back 3000 years (Figure 4.8).

FIGURE 4.8
Skull of the Pech culture and Talgua's karstic caves. *Photo courtesy of www.fotopaises.com.*

GEOMORPHOLOGY OF COPAN VALLEY

Copan is a narrow river valley in the northwest of Honduras, near the border with Guatemala. From a topographical point of view, the studied sector is presented as a series of worn and used mountains whose average altitude is around 3280 ft. Some tributaries of the Copan River are the **Gila** and **Sesesmil** Rivers that run through gorges and valleys that communicate with other areas of the sector. The most important inhabited centers in the sector are Santa Rita de Copan and Copan Ruins.

From a structural and tectonic point of view, the sector is characterized by the outcrop of two major lithologic units. The first unit is the Padre Miguel Formation from the Middle Tertiary, composed of volcanic rocks ranging from rhyolites to andesites to basalts. The second unit belongs to the Valle de Los Angeles group and is formed by sedimentary rocks like limestone, shale, and sandstone. The whole ensemble rests on older sedimentary rocks (Lower Cretaceous Group of Yojoa and Jurassic Honduras Group), which in turn rest on the Paleozoic basement known as Cacaguapa schists. This section is bordered by the Jocotan tectonic fault, which forms part of the tectonic Motagua-Polochic ensemble, and which is why the area has high seismic activity (Figure 4.9).

The Copan River eroded a deep valley, in part exploited by the Jocotan fault and the profile slopes of the river, which have torrential behavior during flood time, charring large blocks of stones and exposing various fluvial terrace levels as well as alluvial fans from its torrential tributaries.

In the Santa Rosa de Copan ruins sector, the river has left evidence of five terrace levels, which can be cyclical, associated with large Quaternary climatic changes. It is worth mentioning the outfall fan, built by the Sesesmil River, where the city of Copan's ruins sit. The alluvial fan has a genesis that involves at least three identifiable episodes.

The first and oldest is a powerful collection of coarse alluvial boulders located on the west bank of the Sesesmil River, which is highly eroded, explaining the unevenness of the streets of the city.

The second alluvial episode is where the Copan archeological ruins (mid-Quaternary) are located, corresponding to an alluvial fan indicating that the Sesesmil River changed course (locked at the mouth of the valley by the oldest cone deposit). The Copan River notched the second cone, taking a bayonet course on the northern riverbank at the foot of the pyramids. It is probably oriented and favored by the north-northwest-south-southeast fault that cuts the river. And it is cone-shaped, forcing the reinforcing of this margin with fluvial support works, made in 1980 by a Japanese mission. The Mayans probably

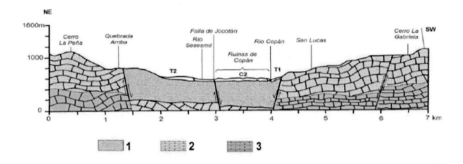

Geomorfología del Valle del Copán

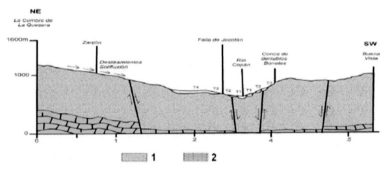

FIGURE 4.9

Two cross sections of the Copan Valley: 1—Tertiary volcanic; 2—Cretaceous sedimentary; 3—Jurassic-Cretaceous sedimentary. *Photo interpretation by J.P. Bergoeing, 2000.*

knew about the problem of catastrophic floods caused by the Copan River, with continuous and annual river boulders transporting; however, it is possible that the river has also been responsible for the destruction of their irrigation channels.

The third, more modern (Holocene) alluvial cone, through which the Sesesmil River runs today, has forced the Copan River to deviate south by drawing a big arch, which limits the same cone. The current C2 and C1 cones (see Figure 4.9, geomorphologic map) have modified the current course of the Copan River, forcing it to describe meanders despite its strong force or competence, favored by the slopes and the flow rate (Figure 4.10).

The Sesesmil River, the main tributary of the Copan River, presents a deep thalweg as a consequence of a fast undermining sector by adopting the shape of a river canyon on its margins. This has forced the river to course through deep gorges in sectors where the structural geological component behaves like a horst by the effects of neotectonics. Hence the Sesesmil River digs its course,

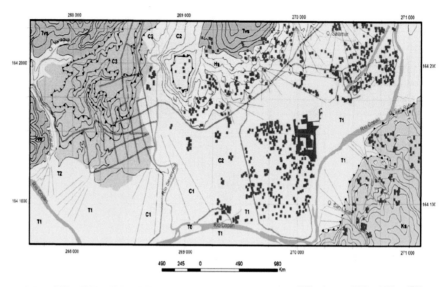

FIGURE 4.10

Sesesmil River cones and geomorphology of the Copan area: 1—HS actual; 2—Terraces and fluvial fans; 3—T2 Fluvial terrace, Holocene; 4—C1 Alluvial modern cone; 5—C2 Intermediate alluvial cone; 6—C3 Old alluvial cone; 7—Tvs. Tertiary volcanic and sedimentary strata; 8—Ks. Cretaceous sedimentary strata; 9—Terrace edges; 10—Solifluction and landslides; 11—Actual debris cones; 12—Hydrology framework; 13—Contours; 14—Copan Mayan ruins; 15—Urban areas *J.P. Bergoeing, 2000.*

favored by one major active fault. Another important aspect of the area is the strong erosive dynamics due to the steep slopes and erosion slopes, where mass landslides and generalized solifluction evolve. This phenomenon has increased because of roads and deforestation.

The field research has made it clear that the morphodynamics of the area of Copan are closely associated with neotectonics, and with the large climatic changes of the Quaternary period that contributed to the modeling of the fluvial terraces on the riverbanks as well as to the constitution of the alluvial cones, which are the reflection of large river deposits (rhexistasy) associated with periods of calm (biostasy).

CHOLUTECA PLAIN

The Choluteca plain forms the Pacific lowlands. It is a narrow alluvial strip where small volcanic cones associated with Central American Pacific

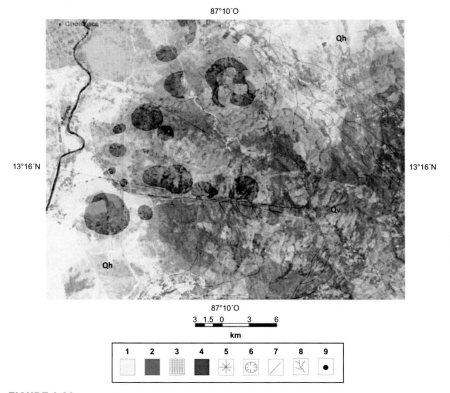

FIGURE 4.11
Choluteca's field sector volcanic geomorphology, Honduras. Cinder cones belonging to the Holocene.
1—Sedimentary, Holocene (QH); 2—Volcanic, Upper Pleistocene-Holocene (QV); 3—Urban areas;
4—Volcanic domes; 5—Volcanic cones; 6—Craters; 7—Tectonic faults; 8—Hydrological network;
9—Villages. *Photo interpretation by J.P. Bergoeing, 2011.*

Quaternary volcanism emerge. The plain is powered by the flooding of the Choluteca and Nacaome Rivers, ending in a vast mangrove swamp (Figure 4.11).

Choluteca Volcanic Field

East of the Gulf of Fonseca, abandoning the mangrove swamps, extends the vast sedimentary plain formed by the Choluteca River during the Upper Pleistocene and Holocene. Choluteca City is also located here. Taking the CA3 road, which runs from the Nicaraguan border to Choluteca, it is possible to find a field of domes and cinder cones a few miles south of the city, in El Palomar sector. They are probably of gas maar origin, and they do not exceed 984 ft in altitude, on average. They are probably a recent demonstration of volcanic activity from the Holocene. Going toward the southeast, in El Obraje, Rio Arriba, and from El Entumido to El Triunfo sectors, a basaltic volcanic complex known as El Entumido volcanic formation emerges.

GULF OF FONSECA VOLCANIC ARCHIPELAGO

This volcanic archipelago is constituted by **Zacate Grande** and **El Tigre** (the biggest island), and by smaller islands like **Garrobo**, **Coyote**, **Violin**, **Inglesa**, and **Conejo**. The last one is disputed with El Salvador. This insular area is composed of a series of active volcanic cones from the Upper Pleistocene that emerged from the sea and are the most recent manifestation of this sector's volcanism, associated with the neighboring volcanism of El Salvador and Nicaragua (Figure 4.12).

The Gulf of Fonseca comprises a 269.11 mi² area that is mostly populated by a great diversity of mangroves that protects a diverse marine fauna, as well as birds and reptiles. The mangroves found here are *Rizophora* sp., *Avicenia* sp., *Laguncularia* sp., and *Conocarpus* sp., which are widespread species in this sector. The Gulf of Fonseca is shared by El Salvador, Honduras, and Nicaragua. The gulf leads to several major rivers that have created one of the most populated mangrove deltaic areas on the Central American Pacific Coast. Among them, from north to south, are the **Amatillo River**, which empties into Bocana del Pecho in La Union's gulf; **Goascaran River**, bordering El Salvador with

FIGURE 4.12
Perfect volcanic dome of El Tigre, dominating the Gulf of Fonseca. *Photo courtesy of Honduraslaboral. org, 2007.*

FIGURE 4.13
Zacate Grande Island is a volcanic Holocene island, the biggest one of Honduras in the Pacific, with a perfect cone and surrounded by mangroves. *Photo courtesy of Honduraslaboral.org.*

Honduras; **Nacaome River**, which flows into Chismuyo Bay; **Choluteca River**; **San Lorenzo River**, where Henecan port stands; and **Sumpile River**. Finally, **Negro River** and the **Gran Estero Real** make up the southern part of this sector, belonging to Nicaragua (Figure 4.13).

CHAPTER 5

The Geomorphology of El Salvador

El Salvador is a volcanic country that emerged during the Tertiary, where tectonic plates are predominant. Five tectonic west-northwest-east-southeast parallel axes cross the country, orientated from the mountain border with Honduras until the coastline. The first axis is located south of the north volcanic range. The second axis is located in the Matapan sector, north of the country. The third axis crosses the country in its central part, where the main volcanic spots emerge. The fourth axis is situated 16 mi off the Pacific Coast, with strong seismic activity. The fifth tectonic axis corresponds to the Central American abyssal pit, which reaches depths greater than 9842 ft with the presence of submarine volcanic cones. Geomorphological landscapes in El Salvador are the natural consequence and prolongation of the same units found in Guatemala. However, the main regions are diversified into four major units:

1. **Northern mountain range bordering Honduras**
2. **Interior valley**
3. **Quaternary volcanic range**
4. **Pacific coastline**

CONTENTS

NORTHERN MOUNTAIN RANGE BORDERING HONDURAS

The northern mountain range marks the northern border between Honduras and El Salvador. It runs from the Metapan to the Coascoran River. The maximum altitude—8956 ft—is reached at Cerro Pital. The mountain range is interrupted by the Lempa River passage, which feeds the Cerron Grande reservoir and then continues toward the east, marking the border with Honduras before going south, leading to the Pacific. The northern mountain range is characterized by Tertiary volcanic rocks (Late Miocene). They are of acidic character (dacites to rhyolites). Near Metapan and Chalatenango, there are intrusive rocks (granite to diorite) that have produced contact metamorphism with marine sedimentary rocks. The ensemble is highly eroded, which explains the emergence of intrusive rocks during the Tertiary.

51

Geomorphology of Central America. http://dx.doi.org/10.1016/B978-0-12-803159-9.00005-4

INTERIOR VALLEY

The interior valley is a vast depressed sector (graben), located south of the mountains bordering Honduras. It runs from Nueva Concepcion through Cerron Grande dam, San Ildefonso, and Hato Nuevo, and ends in La Union Bay (Gulf of Fonseca). The graben is bordered on the south by the Quaternary volcanic range.

QUATERNARY VOLCANIC RANGE

El Salvador has about 23 active volcanic structures, which are distributed along the country following an east-northeast-southwest-south tectonic weakness line. The volcanic range starts in the Ahuachapan Department, which features the volcanic highlands formed by Santa Ana and Izalco's volcanic cones and **Coatepeque caldera**.

Santa Ana Volcano

The Santa Ana or **llamatepec** volcano reaches 7811 ft high and has a 1640.4 yd diameter crater. It is a basaltic-andesitic stratovolcano whose activity is almost Strombolian and phreatomagmatic. It also has continuous fumarole activity. Its last eruption dates back to 1992 and was characterized by cinder emissions. Next to the Santa Ana emerges the **Izalco** volcanic cone, which is a smaller (6463 ft) but similar stratovolcano. It has registered 34 historical eruptions from 1771 to present, so it is considered a very active volcano. Both volcanoes dominate the western **Coatepeque Lake**, which is a collapsed caldera that rises 2447 ft from sea level and is composed of pyroxene andesite. It is an old structure dating from the Lower Pleistocene, also composed of acid pyroclastic flows. However, it has recently presented Holocene activity, as evidenced by ash deposits and the volcanic cone of **El Bigote** north of the caldera's rim (Figure 5.1).

FIGURE 5.1

Santa Ana's volcanic crater dominating the Coatepeque collapsed caldera, formed by granulite andesites of the Lower Pleistocene and an andesitic post-collapse dome inside the lake. *Photo courtesy of Federico Trujillo, Fundacion Coatepeque.*

Santa Ana's volcanic mountains start west of **Las Ninfas** collapsed caldera and continue toward the east with a series of volcanic edifices individualized as **Cerro Cachio** (6046 ft), which has a northern overmouthed crater, and **Las Ranas hill** (5954 ft), which was formed by basic to intermediate lava flows and pyroclastic rocks dating from the Upper Pleistocene. Farther south, along the same west-east line, emerge the more recent cones of **El Aguila** (6561 ft) and **Los Naranjos** (6433 ft), which together with the **Izalco volcano**, date back to the mid-Holocene (6000 years ago).

San Salvador Volcano

West of Sonsonate and toward the San Salvador volcano extend Miocene volcanic highlands (Balsamo Formation), on which are some old volcanic edifices that were formed in the Pliocene and highlight **El Mojon cone** (4101 ft). From there starts a series of east-west cones formed by basaltic lava spills, gradually announcing the San Salvador volcanic area. To the east, and separate from the first set, emerge small volcanic highlands where **San Salvador (Quetzaltepec)** volcano rises 6125 ft and forms a compound cone, presenting Strombolian activity and andesitic-basaltic emissions. It has been studied by several volcanologist missions, and its lava flows have been dated back to 590 AD. It is a volcanic edifice of the Upper Pleistocene, built on a former explosion caldera. To the southeast, it is possible to observe **Buena Vista's** recent volcanic cone (3937 ft) that threatens the Nueva San Salvador urban area. San Salvador stratovolcano skirts have hazardous landslides and lahars' avalanche deposits. San Salvador city is built between the San Salvador volcano slopes and the **Ilopango collapsed caldera**. In this space there are a series of volcanic structures, of which the most important is the basaltic-andesitic **San Marcos volcano** (4265 ft) from the Pliocene (Balsamo Formation), as well as a series of small calderas surrounding the capital suburbs.

Ilopango Collapsed Caldera

Ilopango collapsed caldera is situated east of the San Salvador volcanic complex, measuring 6 mi 1469.7 yd long by 4 mi 1708.9 yd large and 754 ft deep, covering 27.799 mi². Its rim rises 1437 ft above sea level, and its lavas are of dacitic type. A dome appeared between 1879 and 1880 in the Lake Islas Quemadas Islands, showing that this volcanic focus is active and can return to build a post-collapsed important cone of the same dimensions as the former one. The caldera collapsed in the Lower Pleistocene but has annular structures on its edges, which are post-collapsed small explosion calderas. The edges are formed by pyroclastic material (Cuscatlan Formation). Ilopango caldera suffered its most recent Plinian eruption in 430 AD, with lava flows of rhyolitic type. Ashes covered a surface exceeding the current country of El Salvador, leaving pyroclastic deposits and mostly ignimbrite, covering the entire metropolitan area of San Salvador (Figure 5.2).

FIGURE 5.2
Ilopango collapsed caldera. *Aerial photograph courtesy of Roberto Protti, 2011.*

These volcanic deposits are called Tierra Blanca Joven (Young White Ground) because of their important thickness and their position close to the surface of the ground. The homogeneity and isotropic characteristics of the ignimbrite are decisive in the mechanism of the breakdown of slopes consisting of such materials, which collapse, leaving new faces for the slopes with near vertical angles. By this very fact, volumes collapsing are small, but sufficient to cause damage to adjacent homes.

San Vicente Stratovolcano (1 mi 626.26 yd)

The San Vicente stratovolcano, also known as **Chichontepec** or **Las Chiches**, is the second highest volcano in El Salvador. It has two craters, and its slopes are covered with dense tropical vegetation that reaches the top. On its northern and western flanks, a fissure of 896.76 yd allows the rise of fumaroles and hot springs as well as mud springs. The most violent eruption dates back 1700 years, and parts of the volcano show mass landslides on the slopes. We can assume that it was formed in the Upper Tertiary (Miocene-Pliocene) because of the dated basaltic lavas covering the western flank. However, the current cone (2624 ft high) is composed of Upper Pleistocene basaltic-andesitic lavas (Figure 5.3).

San Miguel or Chaparrastique Volcano

San Miguel or Chaparrastique volcano is a stratovolcano with a perfect cone rising 6988 ft, whose emissions are basaltic lava flows. The cone sits isolated from the **Chinameca Range**. Its crater has a diameter of 1093.6 yd and presents adventitious cones near its top. It is one of the six most active volcanoes

FIGURE 5.3
San Vicente volcano with twin cones. *Photo courtesy of Roberto Protti, 2011.*

in El Salvador, with about 26 eruptions in the past 300 years. In 2002, it expressed cinder-type activity. Since 1530, the year of the foundation of the city of San Miguel, this volcano has had eight large basaltic lava flow emissions. In 1762, the most important lava flow reached the city of San Miguel. The volcano also emits ash, gases, and hot mud through its central crater. North of Chaparrastique volcano emerges the **Chinameca stratovolcano**, reaching 4265 ft above sea level; the crater is in fact an explosion caldera of 1 mi 427.23 yd diameter, known as **Laguna Seca El Pacayal**. In addition, to the west it has an adventitious cone, **Cerro Limbo**, higher than the caldera's rim. The fumaroles of this volcano are subject to a geothermal program project (Figure 5.4).

Usulutan Volcanic Complex
Southeast of the Chaparrastique volcano stands a volcanic complex formed by the **Usulutan volcanic cone** (4753 ft), whose summit is highly eroded and whose crater of 1421.7 yd diameter is wide open to the east. **Tecapa volcano** (5226 ft) has three well-formed craters, with a central crater containing a pluvial lake. **El Taburete** (3845 ft) and the ensemble of **Santiago de Maria cones** make up a much-eroded massif with adjacent parasitic cones, of which the most recent is the **Oromantique Holocene cone**.

Conchagua and Conchaguita Volcanoes
The Conchagua and Conchaguita volcanoes are located on the eastern end of the country, bordering the Gulf of Fonseca. This unit was studied by Czech researchers (Rapprich et al., 2010). The **Conchagua volcano** is a volcanic

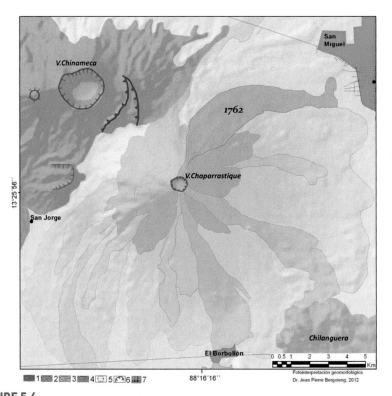

FIGURE 5.4
Geomorphology of Chaparrastique volcano: 1—Chinameca's volcanic area; 2—1762 lava flow reaching San Miguel; 3—Old lava flows; 4—Recent lava flows; 5—Craters; 6—Edges of calderas; 7—Urban areas. *Photointerpretation by J.P. Bergoeing, 2012.*

solid cone compounded by two major cones: **Santiago** or **Ocotal** (4019 ft) and **Banderas** or **Cristobal** (3792 ft). The basaltic **Conchaguita** stratovolcano emerges in the Gulf of Fonseca as an island, rising 1804 ft above sea level. It is the most active sector of this volcanic complex, presenting Strombolian eruptions. In 1892, it issued basaltic lava flows. The geologic history of this complex has been studied by means of radiometric data, with results showing that the oldest rocks are acids (ignimbrites). The volcanic complex emerged during the Miocene, with lavas dating back 12.5 million years in the sector of **La Union** massif. The upper andesites of the **Pre-Conchagua** and **Pancha Juana** sectors give ages of 1.6 million years, and the modern cones of Ocotal and Banderas formed 150,000 years ago. The Conchaguita volcano is associated with lower andesite lavas, dating back 8.4 million years (Pilon Formation). Conchagua's colossus, the archipelago, is formed by outcrops of the volcanic island cones

FIGURE 5.5
Conchagua volcano seen from the top and volcanic archipelago in the Gulf of Fonseca. *Photo courtesy of elsalvadoreshermoso.com, 2010.*

of **Zacatillo**, **Martin Perez**, **Conchaguita**, and **Meanguera**. All these volcanic islands are contemporary with the Conchagua volcanic complex, and can be considered adventitious cones (Figure 5.5).

EL SALVADOR PACIFIC COAST

El Salvador's littoral zone is characterized as the Guatemalan coast by a rectilinear path, and is populated by mangroves in some places. However, the most notable accidents in this sector are in Canton de Puerto Viejo, characterized by a series of volcanic outcrops dating from the Miocene (Balsamo Formation) that determine the coast's route, giving it a convex shape. On the other hand, combined river sediments from the **Grande de Sonsonate** and **Banderas** Rivers have contributed to the construction of a delta (Figure 5.6).

Farther to the east of this sector, the coast turns into a series of small reefs as a result of the extension of the huge volcanic cone of the **Mojon-Comasagua** volcano system, formed during the Late Miocene. The central part of the Salvadorian coast, opposite the city of San Salvador and Ilopango caldera, is conditioned by the construction of the huge alluvial fan of the **Jiboa River** complex, and farther east by a large deltaic cone formed by the **Lempa River**, which includes **Jiquilisco Bay**. Actually, Jiquilisco estuary is the combined result of the contributions of the Lempa River associated with the Pacific littoral stream drift that has regulated the coastline and created the vaster mangrove area of

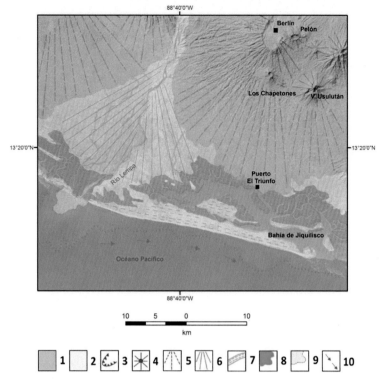

FIGURE 5.6

Geomorphology of El Salvador Pacific Coast and Lempa River delta: 1—Middle to Upper Pleistocene volcanic deposits; 2—Holocene fluvial sedimentary deposits; 3—Explosion caldera; 4—Active volcanic hot spots; 5—Volcanic slopes identified in cones; 6—Alluvial delta fan formed by the Lempa River; 7—Coastal Flandrian littoral cords; 8—Mangrove area; 9—Colloidal suspension sedimentary area of the Lempa River in the sea at its mouth; 10—Direction of the littoral drift. *Satellite photo interpretation by J.P. Bergoeing, 2013.*

FIGURE 5.7

San Salvador, capital of El Salvador, in the evening. At the bottom rises San Vicente volcano. *Photo courtesy of urb.al 3.European Union, 2009.*

El Salvador. The coast is again conditioned by the Miocene volcanic outcrops, from the **Punta El Amatillo** cape to the **Punta Amapala** cape, where coastal cords are narrow and jerky by reefs of volcanic origin. From the Punta Amapala cape to the Gulf of Fonseca, the Conchagua volcano structure determines a coastal path of strong volcanic reefs, which only diminish near **La Union** Port. From this point extends one of the vastest mangroves in the country, from **Bocana El Munio** up the Honduras border, where it continues (Figure 5.7).

CHAPTER 6

Geomorphological Landscapes of Nicaragua

The border between Honduras and Nicaragua is a geologic boundary between the nucleus Central America and the Isthmus Central America. In fact, the border with Honduras cut by **Coco River** cuts the Paleozoic formations, where ridges are oriented south-southwest-north-northeast. It includes the departments of **Nueva Segovia**, **Jinotega**, and part of **Zelaya**, where there are outcrops of Paleozoic, Mesozoic, and Tertiary sedimentary and intrusive rocks (Figure 6.1).

THE PALEOZOIC MOUNTAINS

The Paleozoic mountains are a succession of deformed and folded strata comprising two sections: a foreland in the north, and a geosyncline in the north-northwest that extends to **Mosquitia** territory. The sector was also affected by the Paleozoic orogenesis. During the Mesozoic, the sector was eroded and later invaded by the sea. It emerged again during the Laramian orogenesis, accompanied by intrusive granite that emerged in Nueva Segovia. However, the Mosquitia territory remained under the sea until the Tertiary.

THE CARIBBEAN PLAINS

The Caribbean coastline is characterized by a series of coral reefs constituted by the **Miskito Cays** and **Maiz Islands**, which mark an extensive marine platform. In the continental sector, the littoral is formed from the last foothills of the central volcanic mountain range, featuring a vast Quaternary alluvial plain characterized by flooded alluvial soils and largely covered with a tropical rain forest. The most important sector is located in the Mosquitia basin formed by the Coco River, where Tertiary sediments can reach 16,440-ft thicknesses. The coastline has been divided into two sectors: Northern Caribbean and Southern Caribbean.

Northern Caribbean and Southern Caribbean

The Northern and Southern Caribbean are separated by the **Grande River**. The **Port of Puerto Cabezas**, the second largest of the Caribbean coast, is located in the northern sector. The Northern Caribbean is a vast plain that develops from

CONTENTS

61

Geomorphology of Central America. http://dx.doi.org/10.1016/B978-0-12-803159-9.00006-6

FIGURE 6.1
Granada, Nicaragua's old Spanish Colonial town, on the edge of Lake Nicaragua (bottom). In the center rises the cathedral. This town is situated at the foot of Momotombo volcano. *Photo courtesy of gotravelfurther.com.*

the Coco River bordering Honduras to the Grande River, draining a multitude of rivers descending from the central range of the Departments of Jinotega and Matagalpa. South of the Grande River, the plain is modified in its coastal path by a series of littoral lakes. These are the product of the drainage of the numerous rivers arriving here, as well as the littoral drift that has been aligning the coastline, forming coastal cords that are supported in some cases with preexisting coral reefs (Figure 6.2).

FIGURE 6.2
The town of El Rama on Nicaragua's Caribbean coast, where the last hills of the Maribios Tertiary range are found, and Escondido or Rama River is a fluvial transport way. *Photo courtesy of Julio Lopez, 2009.*

Here stands the **Port of Bluefields**, the chief port of the Caribbean coast of Nicaragua. **Kirinwas** and **Grande de Matagalpa** Rivers drain their waters from its headstreams in the Boaco and Chontales Departments. They have also created a vast alluvial plain, which is complemented with the confluence of the **Wawashan** River. Finally, the navigable **Rama** River allows communications between the Port of Bluefields and the mainland. When arriving in the floodplain, most of these rivers describe a shape of multiple meanders before flowing directly into the sea or through the aforementioned lakes. The meanders are a demonstration of the loss of the rivers' competence, amended only with seasonal flooding.

THE CENTRAL MOUNTAIN RANGE

Tertiary volcanism had already started by the end of the Cretaceous period (65 million years ago), allowing the ascension of the Isthmus Central America, a trough string of volcanic islands that are welded to each other, reaching a climax during the Miocene. However, many of these Tertiary volcanic structures lie on Paleozoic and Mesozoic sediments. It was during this period that the collision between the tectonic Cocos plate and Caribbean plate occurred, creating an anticline in the central sector. This was accompanied by tectonic faults and fissures, through which eruptions of ignimbrites and pyroclastic rocks occurred, marking the end of Tertiary volcanism and creating the Matagalpa structural ignimbrite plateau area. It was the subduction zone push that gave origin to a subsiding area at the foot of the central range, permitting the rise of the new **Maribios volcanic range** during the Quaternary.

THE CENTRAL DEPRESSION

The graben of Nicaragua extends from the Gulf of Fonseca to the Costa Rican Caribbean northern plains, following a northwest-southeast direction. It is limited in the east by the anticline of the central volcanic range, and in the west by the Rivas anticline. Its origin dates back to the Miocene, and it was completed during the Quaternary period as a result of the collision of the Cocos and Caribbean plates. The depression is buried by recent volcanic lava flows and alluvial deposits, covering about 6561 ft thickness, and Lake Managua and Lake Nicaragua remain in the depressed or graben area.

MARIBIOS VOLCANIC RANGE

The Quaternary volcanic transition from El Salvador and Honduras to Nicaraguan volcanism was done through the remains of an explosion caldera, which correspond to the ancient **Cosiguina volcano** dominating the southern

Gulf of Fonseca. It is a solitary volcanic edifice built on a basaltic base. The volcanic summit, which once stood 10,826 ft above sea level, was reduced to a caldera rim that today rises 2814 ft above sea level as a result of the resounding explosion of January 20, 1835. The ashes emitted by this eruption arrived in Mexico, Colombia, and Jamaica in a radius of 1553 mi. The bottom of the explosion caldera, surrounded by deep walls, plunges more than 1640 ft and is occupied by a pluvial lake of blue waters. The volcanic outer walls are occupied by dry, dense, tropical vegetation. The coastal part presents a vertical littoral cliff of 328 ft as well as stratifications of basalts overlapping lava flows, alternating with ashes, pyroclast, and andesitic lava flows. The east volcanic flank also presents an eroded adventitious cone (Figure 6.3).

From this point, separated by the **Nagrandana plain**, the **Maribios range** rises about 49 mi to the interior, formed by a series of active volcanic cones. The Maribios range has a series of volcanoes that can reach up to 6560 ft above sea level, a product of a northwest-southeast tectonic fracture cut by north-northeast-south-southwest fractures, by whose intersections magmatic flows arise. It is a calcareous-alkaline volcanism that has formed stratovolcanoes with the emission of acidic tuffs (Tournon, 1984). The first set consists of the volcanic complex formed by the **Chonco** volcanic cone, followed by **San Cristobal** and **Casita** or **Apastepe** volcanoes. The Casita volcano fails to show signs of historical activity and is seated on a more ancient structure, probably from the Middle to Upper Pleistocene. Therefore, the summit is intensely eroded.

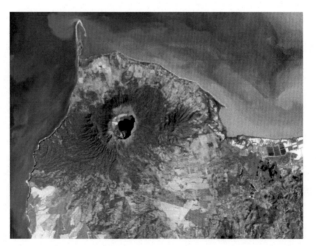

FIGURE 6.3
Cosiguina stratovolcano, Nicaragua. From its 938.32 yd above sea level, it dominates the southern entrance of the Gulf of Fonseca. On the right is a small secondary adventitious cone. *NASA Earth Observatory satellite image, 2007.*

San Cristobal Stratovolcano

At 5725 ft above sea level, **San Cristobal** is the highest peak of this volcanic complex and of the country. Contrary to Cosiguina, this volcano presents intense activity. It has three aligned east-west cones. The main cone is bisected at its summit, consisting of three concentric craters. Its recent eruptions are characterized by pyroclastic flows and ash emissions. The oldest historically registered eruption dates back to 1685, and was useful to buccaneer William Dampier for locating and plundering the city of Leon. It was an extremely violent eruption that framed the minds of the city's inhabitants. Its eruptions are of basaltic-andesitic lava flows, tephra, ash, and pyroclast deposits (Figure 6.4).

Chonco volcano (3625 ft) is coalescing with San Cristobal. It issued olivine basaltic lava flows during the Holocene, and small dacitic domes are present at its base. **Casita volcano** reaches 4609 ft high, and is located east of San

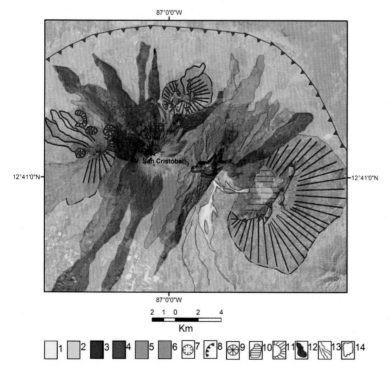

FIGURE 6.4

Geomorphology of the San Cristobal volcanic complex: 1—Holocene sediments; 2—Old lava flows, Upper Pleistocene; 3—Recent lava flows, Holocene; 4—Very recent lava flows, current; 5—Chonco volcanic area; 6—La Pelona volcanic area; 7—Craters; 8—Collapsed calderas; 9—Dacitic domes; 10—Structural volcanic plateau (bottom of caldera); 11—Volcanic cones; 12—Lahar flows; 13—River wash; 14—Volcanic area. *Satellite photo interpretation by J.P. Bergoeing, 2013.*

FIGURE 6.5
San Cristobal volcano's Plinian eruption of September 8, 2012, with emission of volcanic bombs, sulfur dioxide, and ashes that reached 3 mi high. The eruption calmed down 24 h later. *Photo courtesy of Gladys Gomez, 2012.*

Cristobal's crater. It is the structural center of this volcanic complex, and it has issued andesitic-basaltic lava flows, tephra, and volcanic slag, covering its slopes (Figure 6.5).

Farther east emerges **La Pelona collapsed caldera**. It is an old volcanic structure indicating an event prior to the construction of Casita-San Cristobal volcanoes. All these volcanoes are aligned along a northwest-southeast faulting direction. Finally, to the northeast emerges the structure of the **Moyotepe volcano**. It is an explosion caldera, older than the modern complex of Casita-San Cristobal volcanoes. We can also say that the whole complex rests on a mega collapsed caldera structure, which covers the whole volcanic system (Hazlett, 1977). See Figure 6.4.

Telica Volcano

A number of active volcanic cones emerge southeast of San Cristobal. Among them is **Telica stratovolcano** (3470 ft), formed by basaltic-andesitic stratified lava flows. This volcano has presented intense and repeated cinder activity during the last centuries, overshadowing the city of Leon. It has six craters, and at its top has a double active crater. Its last eruption occurred in 2007. In the southwest, it also has a parasitic cone and a series of remnant volcanic cones—the eroded northern **San Jacinto village**—characterized by upwelling hot springs.

Next to Telica volcano is **Rota volcano**, rising only 2742 ft above sea level. According to Incer-Barquero (1980), Rota is the oldest volcano of the Maribios range. No historical volcanic activity has been described, and it has been significantly diminished by erosion. But this does not mean that the volcano is inactive.

Cerro Negro Volcano

Cerro Negro ("Black Hill") volcano emerged in 1850 and grows continuously in altitude with each eruption, currently reaching 2296 ft. Its toponymic feature is characterized by Strombolian eruptions, with abundant pyroclast and ash emissions. Since its formation, it has had 23 eruptions. It is currently a tourist attraction because people practice sandboarding at its skirts or sliding on wooden planks on its slopes, which are covered by volcanic black ash. Next to Cerro Negro rises **Las Pilas volcano** (3543 ft), consisting of several craters. El Hoyo is the most active of the craters, presenting intense fumarole activity. This volcanic group is the result of an ancient volcano known as **Pikachu**, on whose foundation of small caldera the aforementioned cones have been built. Also at its base, the **Asososca Lake** is another remnant of the collapsed caldera (Figures 6.6 and 6.7).

Momotombo Stratovolcano

Momotombo is perhaps the most impressive and beautiful of Nicaragua's volcanic edifices, rising 4653 ft. It forms an almost perfect cone, which is reflected on the waters of **Lake Xolotlan (Lake Managua)**. Next to it emerges the **Momotombito cone** as an island in Lake Xolotlan, which is in all likelihood

FIGURE 6.6
Cerro Negro volcano, one of the most recent and active in the Maribios volcanic range. *Photo courtesy of Oscar Montealegre.*

FIGURE 6.7
Momotombo's perfect volcanic cone and recent lava flow. Xolotlan Lake in the background. *Photo courtesy of Cristina Bruseghini Di Maggio.*

an adventitious cone of Momotombo. The Momotombo summit is devoid of dry tropical vegetation and its crater presents slow, effusive fumarole activity. Its last eruption dates back to 1905, when it issued a powerful north lava flow (Figure 6.8).

It was in 1610, preceded by tremors, that the eruption of the Momotombo caused the destruction of **Leon Viejo**, the old colonial capital, which was built

FIGURE 6.8
Momotombo and Momotombito's volcanic cones dominating Lake Xolotlan. *Photo by J.P. Bergoeing, 2010.*

on the edge of Lake Xolotlan and rebuilt later to the northwest in its current site. Momotombo volcano marks the end of Los Maribios volcanic range and the beginning of a series of isolated volcanic edifices, most of which emerge in the southeast. Two more recent lava flows from the Momotombo crater are oriented northward.

The island formed by the Momotombito cone is also a stratovolcano, rising about 1148 ft above the lake's level. The island was called *"Cocobolo"* by the natives due to the abundance of walking sticks (*Dalbergia retusa*). It is also a pre-Columbian archeological sanctuary, where there are anthropomorphic statues carved in the basalts. It was discovered in 1850 by **Ephraim Squier**, the North American explorer of the Central American area.

Managua Collapsed Caldera

Managua, the capital city of Nicaragua, is an urban area of 67.182 mi^2 that is inhabited by more than two million people. It is seated on a huge collapsed caldera, probably of the Upper Pleistocene or early Holocene, which makes it a very young event. The caldera probably leans toward Lake Xolotlan, with part of it resting under the lake. Inside Managua caldera emerges the post-collapsed **Chiltepe volcanic cone**, rising 1699 ft above sea level. It has two craters: **Apoyeque**, formed some 9000 years ago (INETER, 2002), with a green-water lake at its base, which is obscured by the steep walls of its crater; and **Xiloa crater**, with a blue-water pluvial lake very close to the level of Lake Xolotlan, where fish abound. Both craters are part of the volcanic post-collapsed cone of Chiltepe volcano. If you look at the Shuttle Radar Topography Mission (SRTM) image of Nicaragua, you can see at a glance the vast complex of the Managua collapsed caldera, including Chiltepe volcano and the **Masaya collapsed caldera**, and also **Laguna de Apoyo collapsed caldera**. The study of the generalized Managua stratigraphy column by Rodriguez et al. (INETRER, 2002) suggests that Managua's volcanism is very recent (Upper Pleistocene-Holocene), going back about 29,000 years. However, the volcanic structural base is dated on ignimbrite tuffs and pyroclastic flows from about 870,000 years, which dates it back to the Middle Pleistocene. Post-collapsed volcanism would focus on one greater accident, the Miraflores-Nejapa tectonic fault, where numerous crater structures are inserted following a north-south direction.

The Gas Maars

The term "maar" comes from the Eiffel region in Germany and makes reference to lakes that occupy old volcanic crater episodes. Maars are the product of an explosion between the meeting of a phreatic or underground river and a magmatic rise, which results in the formation of huge amounts of gas rising through fissures and exploding on the surface, creating an explosion

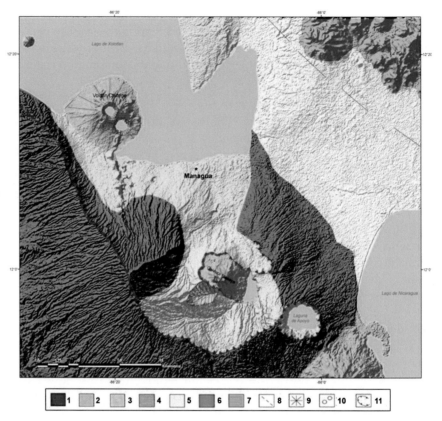

FIGURE 6.9

Managua collapsed caldera. Volcanic evolution: 1—Phase one: Managua volcano; 2—Phase two: Volcano collapse and creation of the Managua caldera; 3—Phase three: Creation of Apoyo and Masaya volcanoes; 4—Phase four: Apoyo and Masaya volcanoes collapse, both calderas form, and Chiltepe volcano rises; 5—QH. Quaternary, Holocene; 6—Vt. Tertiary volcanism; 7—Lakes and ponds; 8—Nicaragua's graben edge; 9—Volcanic cones; 10—Craters; 11—Caldera edges. *Photo interpretation based on SRTM image, J.P. Bergoeing, 2012.*

crater that can become a lake and be powered by the water table. The volcanic field of Kichwambe, south of Lake George in East Africa (Great Rift) is an excellent example (Bergoeing, 2013). This volcanic edifice was generated by phreatomagmatic eruptions from a crater that sits below the original topographic surface of the ground. Also, tuff cones are produced by hydromagmatic eruptions of lower energy, presenting greater lift than the rings of tuffs and products that have less lateral dispersion. The tuff rings are related to hydromagmatic eruptions of high energy, where basal waves are generated. Stopping these waves creates annular deposits surrounding the explosive depression (Figure 6.9).

Miraflores-Nejapa Fissural Area

Gas maars or volcanic gas explosion craters also are present inside the city of Managua. It is a sector located west of Managua between Chiltepe volcano and **Monte Tabor**, following a north-south **Miraflores-Nejapa** tectonic fissure and volcanic area. The area encloses a series of cones, craters, and tuff rings corresponding to phreatomagmatic eruptions (i.e., areas of explosion by the rise of magma from the magmatic camera of the Managua caldera that is still active). Among these structures are **Tiscapa crater lake**, which was formed about 3000 years ago and is 164 ft deep. The circular crater occupied by **Asososca crater lake** of the same age, with 1312.3 yd diameter and 311 ft depth, is in turn surrounded by a series of tuff cones that separate it from **Nejapa crater lake**, another gas maar or explosion crater. The craters are characterized by being quite flat on top and with the same altitude, surrounded by volcanic tuff cones that occur as isolated dark colored hills. Asososca, Tiscapa, and Nejapa are lakes currently fueled by the phreatic layer of Lake Xolotlan, and therefore present a steady flow of water. Other craters of explosion or gas maars located in the southern fissure have no groundwater supply and therefore are dry. However, the north-south Miraflores-Nejapa fissure could be reactivated by tectonic activity and thereby produce new magmatic arrivals that would lead to a new period of gasmaaric explosions, this time in a densely populated area. Lake Xolotlan has an area of 63.430 yd long by 34.996 yd wide, which covers about 402.32 mi^2, corresponding to the submerged part of the Managua collapsed caldera. The lake is fed by the Sinecapa, Viejo, Pacora, and San Antonio Rivers. The city of Managua pours sewage in the lake, contributing about 32 million gallons per day, but the lake is also a mirror of highly contaminated water.

The lake released its waters through the **Tipitapa River** to **Lake Cocibolca (Lake Nicaragua)**. In the Lake Xolotlan sector, we find **Acahualinca human footprints** printed in volcanic ash sediments, deposited by a violent eruption with a radiocarbon (^{14}C) date of 2120 years ago, marking the passage of about 15 natives who were probably fleeing the volcanic event (Schmincke et al., 2008). By its geomorphology, this event can be associated with gas maar explosions, as we can see from the small craters of explosion of the adjacent sector lake (Figure 6.10).

On Wednesday, October 9, 2013, Angelica Muñoz, spokesperson for INETER, the Nicaraguan Institute for Territorial Studies, issued a "technical alert" for a swarm between 1.8 and 2.9 degrees on the Richter scale occurring under Lake Xolotlan. "Earthquakes caused by local (geological) faulting, which comes from Asososca and passes through Nejapa and Acahualinca, continue in the depths of the lake," said Guillermo Gonzalez, chief executive of SINAPRED, the National System for Prevention, Mitigation, and Attention of Disasters, fearing an earthquake like the one in 1972 (La Prensa de Nicaragua, October 9, 2013).

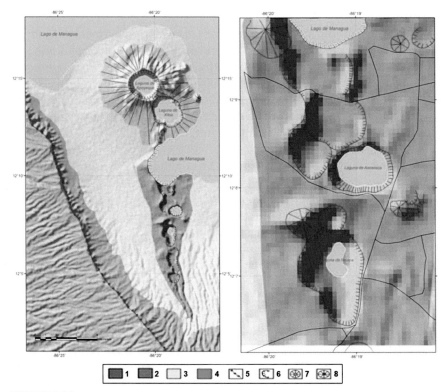

FIGURE 6.10

Miraflores-Nejapa fissure and gas maar area: 1—Sedimentary, Holocene (QH); 2—Volcanic, Upper Pleistocene-Holocene (QV); 3—Urban areas; 4—Volcanic domes; 5—Craters; 6—Tectonic faults; 7—Hydrological network; 8—Villages. Geomorphological photo interpretation by *J.P. Bergoeing, 2012.*

On April 10, 2014, a shallow 6.2 magnitude earthquake struck Lake Xolotlan. Additional earthquakes were registered through April 17. Two persons were killed, and 2300 houses were damaged. The epicenter was determined to be 1 km from Apoyeque Crater Lake of the Chiltepe volcano on the shores of Lake Xolotlan (Figures 6.11 and 6.12).

Masaya Volcano Caldera Complex

The Masaya volcano caldera complex is located southeast of Los Pueblos volcanic plateau. The volcanic group constitutes a collapsed caldera made up of four craters—the active **Masaya**, **Santiago**, **Nandiri**, and **San Pedro**—separate from the main crater, which also shows fumarole activity.

The complex corresponds to the formation of volcanic cones after the Managua caldera collapsed, which evolved into a volcano about 10,000 years ago and subsequently also collapsed, creating a new caldera in the interior of the

FIGURE 6.11
Acahualinca footprints left by a native group heading toward Lake Xolotlan in search of boats, probably fleeing from a gas maar eruption 2000 years ago. Deep tracks indicate a person carrying someone on his back. *Photo by J.P. Bergoeing, 2013.*

FIGURE 6.12
Asososca gas maar crater lake and tuff cone. *Photo courtesy of Dalila Maria Montealegre, 2012.*

Managua caldera that is smaller than the previous one. Inside the caldera, it is possible to observe numerous basaltic lava flows. The two main lava flows (from 1670 to 1772) are directed toward the north and are the most important because they reach Lake Xolotlan. It is in the eastern part of the complex where edges of the collapsed caldera are partly occupied by a pluvial lake known as "Laguna de Masaya." Caldera's rim has a 262-ft free fall. Jaime Incer-Barquero (1980) says that the volcanic complex represents a shield type of volcanism. He attributes its rising to the Holocene (2500 years ago), and the volcanic base is composed of basalts and tephra. During the collapse process of the caldera, it also issued ignimbrites and could have been a violent explosion. The Nandiri crater erupted in 1670 and in 1772. Another eruption with emission

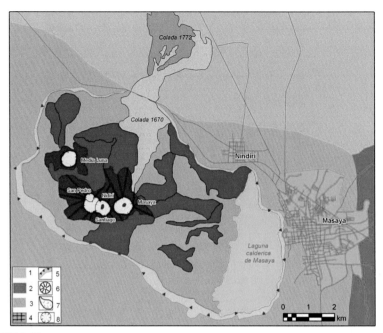

FIGURE 6.13

Masaya collapsed caldera: 1—Managua collapsed caldera area; 2—Apoyo paleovolcano area; 3—Masaya paleovolcano area; 4—Urban areas; 5—Rim of caldera collapse; 6—Volcanic cone; 7—Lava flows; 8—Volcanic craters. *Geomorphological photo interpretation by J.P. Bergoeing, 2013, based on a Google Earth 2012 satellite image.*

of basaltic lava emerged by a fissure of Masaya's cone and a surge of lava flow that was probably from 1772, turning northward and reaching Lake Xolotlan (3 mi). It was the most important flow of this complex in modern times. The 2003 eruption was of ashes and steam, rising to 9000 ft high. The last eruption occurred in 2008. Santiago, the only active crater, emits fumes of sulfur dioxide (Figures 6.13 and 6.14).

Apoyo Collapsed Caldera

Next to Masaya's caldera lies **Laguna de Apoyo**, another collapsed caldera. This one is the most ancient. It is almost circular, rising a diameter of about 2.5 mi and with vertical walls formed by basalts, dacites, tiffs, slags, cinders, and laharic material stratifications. The lake occupies all of the depression and is of singular beauty. It is 577 ft deep, of which 328 ft are below sea level. At its edges, it presents hot springs. The former **pre-Apoyo volcano** collapsed about 23,000 years ago (Espinoza et al., 2008), creating the collapsed caldera, probably contemporary with the formative episode of the Managua caldera (Figure 6.15).

FIGURE 6.14
Masaya's volcanic cone rising in the caldera and pluvial lake. *Photo by J.P. Bergoeing, 2013.*

FIGURE 6.15
Laguna de Apoyo's collapsed caldera structure, probably contemporary with the Managua caldera, taken from Catarina terrace. *Photo by J.P. Bergoeing, 2013.*

Los Pueblos Plateau

Los Pueblos plateau is an extensive volcanic area reaching 3064 ft above sea level, dominating the city of Managua. It gradually descends toward Lake Xolotlan through neotectonic steps. Several rivers carve the Pacific slopes of this plateau, which is the remnant of the ancient Pleistocene edifice of the former Managua volcano that occupied the entire sector of Managua and Masaya and is a silent witness of the moved volcanic activity in this sector of

Nicaragua, whose example is the Masaya caldera located on its southeastern flank. Sequential chronology of the volcanic events can be described as follows:

900,000 B.P. Rising and building of Managua volcano
27.000 B.P. Eruption and collapse of Managua volcano and pre-Apoyo volcano
25.000 B.P. Rising of pre-Masaya volcano
15.000 B.P. Rising of Chiltepe volcano in the Managua caldera
10.000 B.P. Eruption and collapse of Masaya volcano; creation of current caldera
3.000 B.P. Miraflores-Nejapa fissure, where the explosion of gas maars occurs (Acahualinca traces)

From the observation and description of the deposits in the study sector, the conclusion is that the Asososca and Nejapa craters and all those located in the Miraflores-Nejapa fissure are gas maars or explosion craters and tuff rings formed by intense phreatomagmatic eruptions. Eruptions resulted in the emission of pyroclast waves, with explosion breaches the result of hydromagmatic changes in this sector. The Managua area is particularly vulnerable because it is now situated in a densely populated area. In its surroundings, the Masaya caldera is equally worthy of surveillance because an eruption like the one in 1772 would affect a large number of residences. Laguna de Apoyo is also an active collapsed caldera, and could reconstruct its initial cone. Therefore, we must be aware that these episodes will be repeated in time, and take preventive measures. Managua, the capital of Nicaragua, is an urban area today, with 1.5 million inhabitants seated in a huge collapsed caldera that leans toward Lake Xolotlan, with part of it under the lake. Inside the caldera, Chiltepe volcano has two craters, including Apoyeque Lake. If you look at the SRTM image of Managua, you can see at a glance that the Managua caldera is a vast complex that includes Chiltepe volcano, Masaya collapsed caldera, and probably Apoyo caldera. A study done by Rodriguez et al. (INETER, 2002) of the generalized stratigraphic column of Managua suggests that Managua's volcanism is very recent (Upper Pleistocene-Holocene), dating back about 29,000 years. However, the volcanic structural base is dated by ignimbrites, tuffs, and pyroclast flows from about 870,000 years, which brings us back to the Middle Pleistocene. Post-collapse volcanism would focus on one greater accident, the Miraflores-Nejapa fissure, where numerous crater structures are inserted in a north-south direction.

Mombacho Stratovolcano

From its summit reaching 4409 ft above sea level, **Mombacho volcano** dominates the city of **Granada** and the northern area of **Lake Nicaragua**. This volcano began to take shape in the Upper Pliocene and has been building up in violent eruptive phases, testified by its geomorphology. Before the event that caused the summit explosion calderas, this volcano had reached approximately 5500 ft in altitude, constituting a volcanic massif that was wiped out with the bang quarter.

The volcanic summit consists of two explosion calderas reaching almost 1.5 mi in diameter, with vertical walls of 2132-ft depth. The first and most important caldera opens toward the northeast, separated from the second caldera by a thin volcanic wall. This caldera drained a powerful mass of basaltic lava flows that reached Lake Nicaragua, creating **Asese Peninsula**, and leaving 360 basalt islets of all sizes, known as *"Las isletas."* This event probably happened in the Upper Pleistocene, due to volcanic material freshness. Also, this area presents andesitic and dacitic lava flows scattered northwest of its skirts. The inner Asese Bay holds equally small volcanic explosion craters that linger in **Zapatera island.** The second explosion caldera opens toward the southwest. It is more circular and produces equally powerful lava flows (Figure 6.16).

The current summit is also occupied by two recent volcanic craters, from which the last historical volcanic lava flows with pyroclastic effusion. It is likely that the 1570 event flowed north from the main crater, but according to colonial chronicles, it flowed south and destroyed a native village of 400 inhabitants. I believe the most recent lava flow drained northward, emanating from two fresh craters that are in the summit (as seen in satellite images). Colonial chronicles that recount the event may be indicating the area south of the city of Granada. This active stratovolcano contains fumaroles and hot springs, and its flanks are also affected by laminar and mass landslides. In 1574, in the *"Description of the District of the Audience of Guatemala"* (Revista

FIGURE 6.16
Mombacho explosion caldera seen from the southwest and from Apoyo Lake caldera, from which emerged powerful lava flows. *Photo courtesy of Agencia de turismo responsable El Perezoso.*

del pensamiento Centroamericano, 1970) **Juan Lopez de Velasco** makes the following comments: *"Because of an earthquake, the south wall of the volcano* (Mombacho) *collapsed causing an avalanche of mud and stones* (lahar). *The indigenous Mombacho's people were buried dying its 400 inhabitants. **Four Leagues from the city*** (Granada) ***was an Indian village called Mombacho, next to a small volcano to the year 70*** (1570) ***with a very large storm of wind and water that happened at night, and a mud fell to all people who was in it, nobody escape but a single neighbor of the city of Granada called Carvallo and two old Indian women: being six or seven Spaniards with all other Indians buried. On the other hand came out so great tempest of water and stone hurting cacao plantations and cattle farms"*** (Figure 6.17).

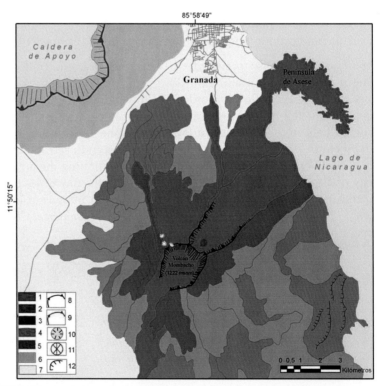

FIGURE 6.17

Mombacho's volcano geomorphology: 1—Recent 1570 lava flow; 2—Lava flows from Pleistocene to Holocene; 3—Lava flows of Upper Pleistocene; 4—Lava flows from Middle to Upper Pleistocene; 5—Lava flows of Middle Pleistocene; 6—Pre-Apoyo volcano area and collapsed caldera from Upper Pleistocene; 7—Volcanic soils for crop use, Holocene; 8—Edge of explosion calderas; 9—Collapsed calderas rim; 10—Recent volcanic craters; 11—Volcanic domes; 12—Erosive slopes. *J.P. Bergoeing, 2013.*

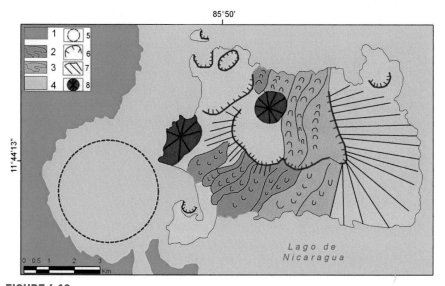

FIGURE 6.18

Zapatera Island volcanic geomorphology: 1—Mombacho volcanic area; 2—Recent lava flows; 3—Old lava flows; 4—Zapatera volcanism area, Holocene; 5—Submerged caldera; 6—Volcanic craters; 7—Volcanic cone; 8—Volcanic domes. *J.P. Bergoeing, 2013.*

Farther south, **Zapatera island** rises 1968 ft above sea level and is a volcanic relic of the Upper Pleistocene, covered by basaltic lava. In the center, there is a lake that occupies an ancient crater covered by rainwater. To the west is a vast circular depression in Lake Nicaragua, the testimony of a former caldera. To the north and west are a number of half-submerged craters bordering the island (Figure 6.18).

Ometepe Island: Concepcion and Maderas Volcanoes

Ometepe island is the result of the coalescence of two volcanic cones covering an area of 106.56 mi², forming a single island. The island unites **Concepcion stratovolcano,** rising 5282 ft above sea level with slow effusive activity, and **Maderas stratovolcano,** rising 4573 ft in altitude, without historical activity registered and with a summit crater occupied by a pluvial lake. This insular set has two perfect volcanic cones. Concepcion volcano was known in pre-Hispanic times as **Omeyatecigua;** it has had eruptions recorded since 1883. In 1957, it issued pyroclastic material, and in 2006-2007, it erupted ashes. The volcano presents a risk of lahar landslides on its southern flanks associated with eruptions. The last eruption, in 2010, was of cinders and sulfur dioxide, and the ash column rose 22,000 ft high (Figure 6.19).

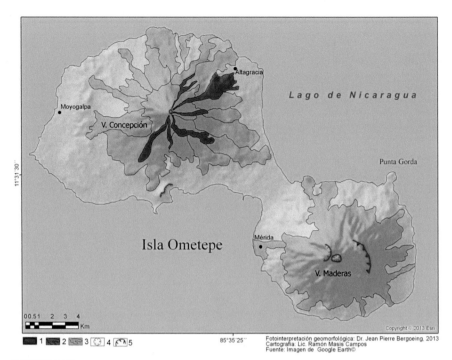

FIGURE 6.19

Ometepe's island geomorphology: 1—Recent lava flows; 2—Pleistocene volcanic base, old lava flows; 3—Volcanic craters; 4—Edge of explosion calderas; 5—Volcanic rim. *Photo interpretation by J.P. Bergoeing, 2013.*

GEOMORPHOLOGICAL CHALLENGE: NICARAGUA'S INTEROCEANIC CHANNEL

Since colonial times, Nicaragua's isthmus has been discussed as an option for maritime traffic, used by travelers since the middle of the nineteenth century. However, the opening of the Panama Canal was a blow to this project, which has not been forgotten but has been postponed for a long time. In 2004, Nicaragua's government revived the project, with the aim of making 250,000 tons of transit vessels, with an approximate cost of $25 billion (25 times the annual budget of Nicaragua). A project of this amplitude cannot be done without foreign capital. The Grand Canal Commission was created by presidential decree in December 1999 and renewed by means of two decrees in 2002 and 2006. In 2012, Nicaragua's National Assembly approved by large majority the research for an interoceanic canal of great depth that would unite the Caribbean Sea with the Pacific Ocean, with an approximate cost of $30 billion (Figure 6.20).

FIGURE 6.20
1885 interoceanic German project by Nicaragua. *Source: Wikipedia/Historia del canal de Panama.*

Feasibility of the Project
The construction of an interoceanic maritime channel raises several problems that need to be thoroughly studied.

1. **San Juan River.** Since the nineteenth century, this path has been chosen many times, but stumbles upon the political-administrative reality that its south bank belongs to Costa Rica. Thus, the San Juan River cannot be used.
2. **Amount of water needed for locks use.** Nicaragua is located in the dry tropical climate zone, with six months of drought per year, making the levels of Lake Nicaragua thrive during this period. Therefore, water from the Pacific Ocean and the Caribbean should be pumped to the lake, leading to a change in the aquatic environment. This would save less than the water used in the locks, to be put in tanks suitable for future reuse.
3. **Proximity of Ometepe island.** With two active volcanoes (Concepcion and Maderas), a prolonged eruption could paralyze lake traffic. The bathymetric measurements conducted in Lake Nicaragua between Las Lajas river mouth (Rivas isthmus) and Ometepe island in a stretch of about 16 mi gives an average of 16-ft depth. The minimum depth required for an interoceanic channel is 88 ft. The lake bottom is composed of basaltic lavas, so dredging would be almost impossible (La Prensa journal, September 2, 2013).
4. **San Isidro dam.** The need to build San Isidro dam in San Juan River will increase the level of flooding of Costa Rican land at the international border.
5. **Commercial competition.** There is sufficient maritime capacity to make commercial competition with the current Panama Canal to implement a third channel and earn about $400,000 for each ship that passes.

Chontaleña Mountain Range

The Chontaleña mountain range is the major obstacle for the future channel. We can limit this range by saying it starts in the south bank of **Grande de Matagalpa River,** and from this sector it is constituted by a set of low ranges that do not exceed 820 ft and are arranged by northwest-southeast orientation to touch the San Juan River. At 4101 ft above sea level, **Cerro Alegre** mountain is the highest point. This mountain range is of volcanic origin and dates back to the Tertiary, about 60 million years ago, reaching its biggest volcanic development during the Miocene, about 15 million years ago. The Chontaleña range divides Nicaragua into two watersheds: one overlooks Lake Nicaragua with the **Mayoles River;** a second one, which is much more developed, is the Caribbean watershed, where **Mico** and **Rama Rivers** are born. The fluviolacustrine slope widens toward the southeast up to the San Juan River mouth, which in San Carlos drains Lake Nicaragua to the Caribbean. Rivers of this watershed that feed Lake Nicaragua are rushing and suffer floods during the rainy season (May to November). Oyate River marks the sector where the Chontaleña range is divided into northern and southern sectors, separated by hills that do not exceed 656 ft high, except in numbered sectors occupied by isolated hills. This natural "window" of the range is that which has been privileged for construction of the canal, due to its low altitude. In 2013, Nicaragua's government signed a contract with Wan Jing, a Chinese businessman, for construction of the interoceanic canal in Nicaragua to begin in 2014 (Figure 6.21).

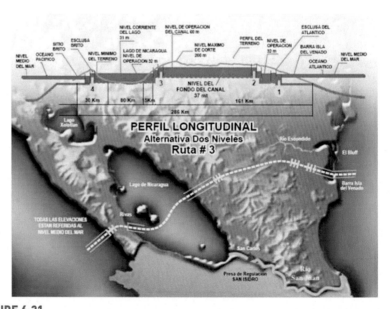

FIGURE 6.21
Chosen route for the construction of Nicaragua's canal. *Published by La Prensa journal on July 31, 2013.*

NAGRANDANA PLAIN AND PACIFIC COAST

The **Nagrandana plain** is a vast region located northwest of Nicaragua's Pacific Coast, bounded on the north by the Cosiguina volcano, on the east by the Maribios Quaternary central volcanic range, and on the south by Lake Xolotlan, marking its natural limits. To the west, it borders the Pacific Coast. The Nagrandana plain is a low area, the logical succession of the volcanic foothills of the Maribios range. It is crossed by many rivers in short journey that irrigate the plain and drain into the Pacific Ocean, constituting vast deltaic areas populated by mangrove lagoons, which grow easily protected by the coastal Flandrian sand ridges. The Nagrandana plain hosts two major urban centers: Chinandega and the former colonial capital Leon. Its fertile lands, paved with volcanic ashes, are dedicated to sugar cane, bananas, rice, and formerly cotton cultivations. Because it belongs to the "dry Pacific" sector, this plain suffers wind erosion during the dry season (December to May).

Northwest to southeast, Nicaragua's Pacific Coast can be divided into five large sectors as follows:

1. **Delta sector with large tracts of mangrove vegetation.** In the Gulf of Fonseca, **Estero Real** is characterized as a very flat and low region in the process of silting by mangrove vegetation. It is a depressed sector where rivers flow into Somotillo's plain. It serves as Rio Negro River's international border with Honduras. The mangrove is the same extension we find in El Salvador and Honduras, extending along the coast of the Gulf of Fonseca.
2. **Active cliff sector.** This sector corresponds to Cosiguina's volcanic cone, forming a cliff undergoing constant Pacific Ocean erosion that leaves vertical walls where different volcanic events that built the Cosiguina volcano are exposed.
3. **Long coastal cords, lagoons, and mangrove coast.** This sector starts in Punta Cosiguina Cape and continues about 55 mi southeast to the mouth of the Escalante River in Nicaragua's Central Pacific sector. This sector consists of narrow, long, almost straight Flandrian coastal cords, ending in coastal curved sand arrows. It is home to four mangrove sectors: Estero Padre Ramos, Estero Aserraderos, Estero de Corinto, and Estero Juan Venado. Corinto's mangrove area is the most extensive, and it is where the country's first port, the Port of Corinto, was built.
4. **Long coastal cords without internal mangroves.** This sector is situated between Escalante River and Astilleros Bay. It represents the nearest coast from the capital city of Managua, which is a popular recreation area. The sector is also characterized by small active cliffs of volcanic origin (usually basaltic). Here, the Flandrian coastal cord sits directly in the preexisting coastline of Los Pueblos plateau. Coastal cords are jerky

FIGURE 6.22
San Juan del Sur, sedimentary folded Tertiary cliff, drawing small sheltered bays. *Photo by J.P. Bergoeing, 2013.*

because of the many rivers that descend from the plateau, forming a radial fluvial pattern.

5. **Little arc bays and beaches area.** This sector corresponds to the Southern Pacific sector of Nicaragua, in the Rivas Isthmus composed by Cretaceous-Tertiary sedimentary rocks (Rivas and Sabana Grande Formations; Dengo, 1973). It is distributed from Astilleros Bay to the Salinas Bay border with Costa Rica. It is characterized by a series of spurs or rocky points that host small, well-sheltered sandy bays, such as Astilleros, Manzanillo, Nacascolo, San Juan del Sur, Anima, and Ostional. Rocky rams also present rocky platforms at their base, with concave profiles by marine erosion. The coast present islets and low funds, a product of the marine erosion that wears away sandstone more easily for the benefit of limestone (Bergoeing, 1987a,b,c) (Figure 6.22).

CHAPTER 7

Costa Rica's Structural Units

Costa Rica, Nicaragua, and Panama form the Isthmus Central America. They are mainly recent volcanic lands, created at the end of the Cretaceous, only about 70 million years ago in geological time. While Costa Rica only has an area of 19,691 mi², it has a wide variety of geomorphological landscapes as a result of its evolution throughout the Tertiary and Quaternary. The large structural units of Costa Rica are distributed from northwest to southeast. In the northern half of the country, units ranging from the Caribbean to the Pacific are as follows:

1. **Nicaragua's Graben Extension Northern Plains**. The plains correspond to the vast basement plain of the Caribbean and go south and southwest, progressively shaping the Guanacaste and Central volcanic range piedmonts.
2. **Plio-Quaternary Volcanic Ranges**. These are the Guanacaste and Central volcanic ranges on a high tectonic position. In this sector, they are the country's topographic and hydrographic axis.
3. **Tempisque River Tectonic Depression**. The depression corresponds to an inlet of the Gulf of Nicoya and constitutes the basal level of Tempisque River. Piedmont deposits brought from the Guanacaste volcanic range (almost ignimbrite) form Liberia's great plateau and serve as a transition from the previous tectonic unit.
4. **Tertiary Mountain Ranges:**
   4A. **Tilaran Mountain Range**. From Upper Miocene-Pliocene volcanism, this range corresponds to the formation of Aguacate.
   4B. **Talamanca Mountain Range**. This range in southern Costa Rica is where the highest point of the country is located (Chirripo Mountain, 12,529,527 ft above sea level). Only part of the Talamanca range is volcanic (Miocene-Pliocene). The base is formed primarily by Tertiary marine folded and faulted series, intruded by granodiorite during the Miocene that constitutes a batholith. From the Caribbean coast to the Pacific, crossing through the northern foothills and the axial zone of the Talamanca mountain region, of difficult penetration and covered by rainforest, the slopes accused

CONTENTS

85

Geomorphology of Central America. http://dx.doi.org/10.1016/B978-0-12-803159-9.00007-8

dissymmetrical sheds of being the shortest and strongest in the Pacific watershed, but broad and well developed in the Caribbean.

5. **The Pacific Coast.** The Pacific Coast is the most developed in the country, with four important peninsulas: Santa Elena, Nicoya, Quepos, and Osa. It is where we find the oldest rocks of Costa Rica.

5A. **Cocos Island in the Pacific**

6. **Caribbean Coast.** The Caribbean coast is divided in two parts: the northern part is formed by parallel coastal ridges up to the Port of Limon, and the southern part is formed by short bays with paleocoral reefs (Figure 7.1).

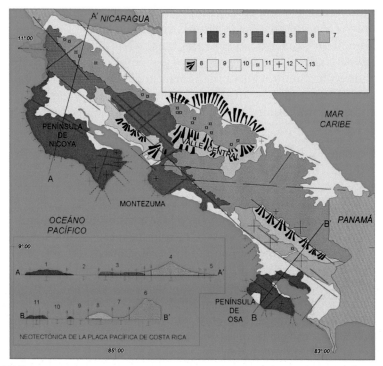

FIGURE 7.1

Neotectonics of Costa Rica's Pacific Coast. Relief units: 1—Peninsula de Santa Elena sedimentary and ophiolitic series (end of Cretaceous and Tertiary); 2—Peninsulas of Nicoya and Osa (Cretaceous-Tertiary); 3—Talamanca range (Tertiary volcanic, sedimentary, and intrusive); 4—Ignimbritic plateau of Liberia (Plio-Quaternary); 5—Tilaran range (Tertiary: Pliocene); 6—Guanacaste and Central volcanic ranges (Quaternary); 7—Coastal range (Tertiary: Pliocene); 8—Big alluvial fans of the Quaternary period; 9—Tempisque's Diquis and Coto Colorado tectonic depressions; 10—Graben of Nicaragua; 11—Sites of Montezuma and Puerto Jimenez (estimated rising of 2-4 m per millennium); 12—Identified raised levels; 13—Tendency to subsidence. Cut A-A': Peninsula of Nicoya, Tempisque depression, Liberia's ignimbritic plateau, Guanacaste range, graben of Nicaragua. Cut B-B': Osa Peninsula, depression of Diquis, Coastal range, depression of Coto Brus, Talamanca mountain range. *J.P. Bergoeing, 2007.*

NORTHERN PLAINS EXTENSION OF NICARAGUA'S GRABEN

Lake Nicaragua (Cocibolca) formed gradually during the Quaternary period. In effect, the ending Tertiary orogeny left folded (anticline) sediments of the Rivas formation (mainly sandstone and limestone) exposed, constituting an isthmus of 13-18 mi wide, which certainly interrupted contact of the southwestern Nicaraguan region with the Pacific Ocean. Inside, Nicaragua's tectonic depression was occupied by a marine origin lake whose waters would become freshwaters over the millennia, thanks to the numerous fluvial tributaries that fed it. Lake Nicaragua was reduced considerably throughout the Quaternary. Today, the lake's dimensions cover an area of 3191.1 mi^2, and its major axis is 100 mi. Bathymetric studies indicate that the average depth is 42 ft, but maximum depth reaches 196 ft to 229 yd at 1422 yd southeast of the island of Ometepe. The lake's volume is estimated at 27,473.9789 gallons. At the beginning of the Quaternary period the lake was salty and marine fauna (the *Carcharhinus leucas* shark) lived there, gradually adapting to the conditions of lower levels of salt due to the surrounding rivers' contributions coming from the Tertiary volcanic foothills (northeast) and Quaternary hills in construction (west and south). During this period, communication with the Caribbean was broad. The intergraben sector extending to the south in Costa Rica started refilling with sediment from the Central and Guanacaste volcanic ranges, with many lahar contributions to the foot of the slopes, along with the construction of powerful alluvial fans that would reach linear dimensions of 24-38 mi to the north.

> The post-orogenic period correspond to the Pliocene and Quaternary characterized by an important development of the faulting system and formation of big blocks that definitely make the major structural units. Particularly, which consequences of tectonic activity in the sector of major laminar faulting, subsidence basins as the Nicaragua's depression will be transformed into real grabens. These depressions were accumulation centers of the erosive elements of emerging Ranges concomitantly with products of volcanic eruptions that occur at that time.
>
> **Butterlin (1977)**

Also, a little less-developed volcanism started in the Tortuguero-Sierpe sector, leaving small isolated mountain ranges in the middle of the flood plain under construction (Tournon, 1972). However, the intergraben sector also began to give signs of rising starting in the Middle Pleistocene. Diluvial rains characterized the Upper Pleistocene period and ended the construction of the powerful alluvial fans, taking a multi-convex shape, thanks to the consistent contributions of the volcanic mountain chains always in construction. The positive neotectonic is a constant that must be taken into consideration from that moment on. During the Upper Pleistocene (i.e., about 200,000 years ago), Nicaragua's southern extension of the lake went about 13 mi to the interior of Costa Rica in an area stretching from Los Chiles to Caño Negro, which was probably affected by the Eemian marine transgression because communication with the Caribbean Sea was still partly open.

Alluvial fans of the Middle to Upper Pleistocene were built from the lahars' lower limit deposits of the volcanic mountain ranges, covering the southern limit of Nicaragua's graben. They extend deeply northward, reaching important linear distances (37 mi). They are mainly made of decomposed ferruginous red clays. Upstream, these alluvial fans contain some volcanic blocks (andesites, tuffs, and basalts) that are several inches in diameter and mixed chaotically within this matrix, which is partially altered in surface. This material has better resisted the general breakdown of the lithic material swept away by river avalanches of this period, which alternates between rhexistasic and biostasic periods (Bergoeing, 1998). The alluvial fans are jerky because of paleochannels and modern channels, giving this sector a multi-convex shape.

Fluviolacustrine Terraces

From the Upper Pleistocene to the Holocene, the various rivers of Lake Nicaragua, including Frio River (Rio Frio), deposited sediments in the bottom of the paleolake, which are of the same genesis clay as the alluvial fans, because these are eroded and the material was transported in suspension by runoff. Between the border of Los Chiles and Nicaragua arises an elongated northwest-southeast fluvial terrace, rising about 32 ft. from a general elevation of 131 ft above sea level. This terrace has a wavy surface, shaped by the contributions of Frio River, which currently flows 11 mi from this point to the north, forming a lacustrine delta in the lake a few miles before the mouth of San Juan River's natural drain of the lake. The top of the terrace consists of decomposed red clays that is leached with whitish spots and whose thickness is about 32 ft. This strata rests on a gray lacustrine clay level (base level) that does not exceed 3 ft thickness, and this second level at the same time rests on a lacustrine white clay level.

The current paleolacustrine level of Los Chiles is therefore situated at 98 ft elevation (Lake Nicaragua's surface level measured by INETER of Nicaragua gave a value of 103 ft above sea level). We can calculate the sector between Los Chiles and the lake as a distance of 11 mi by the fact that Frio River was settled only about 17,000 years ago (if we calculate that sediments deposited in a lake gain approximately 3280 ft every 1000 years). It should be noted that the Flandrian transgression (−6500 years) probably interrupted the sedimentation process; therefore, the sedimentation rate could be much faster and, in this case, fully Holocene. The fluviolacustrine terrace of Frio River leaves a new fluviolacustrine terrace well exposed on the bridge's left bank between Los Chiles and Caño Negro, characterized by 33 ft of leached red clay deposit. It is a regional sedimentation that characterizes the Upala and Los Chiles low sectors. It corresponds to the upper sediments of the lacustrine Los Chiles terrace. At the top of this terrace, Frio River has built a composed fluvial terrace by storing yellow slime-sandy sediments that correspond to the seasonal flooding area, caused by water floods of this river. At the base, almost in contact with the current river, the red clays of the top level rest on a black clay stratum, about 8-12 in., which we can also observe in the current margins of Frio River in Los Chiles. Under this black layer appears a clear silt stratification of lacustrine clay, which represents the base level (Figure 7.2).

Frio River Basin

Frio River originates, strictly speaking, from a number of tributaries born in the Guanacaste volcanic range, between the slopes of **Tenorio volcano** and **Cote Lake.** The basin presents high risk of erosion. Its mouth is formed by a delta not very far from the San Juan River drain. The amount of sediment transported by this water system in the upper part of the basin is demonstrated by the following study: *"The specific sediment amount calculated in Guatuso's station was 667.520 pounds by 0.3861 miles^2/year and in Venado station 405.440 pounds by 0.3861 miles^2/year, of lesser magnitude than the San Carlos River contributions, but with the same implications on turbidity."* (PNUMA-OEA 1997) This study not only demonstrates the amount of current sediment transported in suspension and dilution by Frio River, but also the immediate consequences in sedimentation deposits of the wetlands in

FIGURE 7.2
Los Chiles fluviolacustrine terrace. Superior level: Fluvial deposits with small runoff in a clay caolinitic red matrix. Bottom: Lake deposit stratifications. *Photo by J.P. Bergoeing, 2006.*

the flooded areas, particularly during the summer. It is inferred that during the Middle and Upper Pleistocene in particular, and under a more contrasted climate (rhexistasic and biostasic phases), Nicaragua's tectonic depression has quickly filled by decreasing the lake surface and transporting its coastline northward.

PLIO-QUATERNARY VOLCANIC RANGES

Guanacaste Volcanic Range

The Guanacaste volcanic range belongs to the Quaternary. It is a continuation of the northwest-southeast fracture in Nicaragua. It has been characterized by ignimbrites' acidic eruptions that have established a 328-524 ft plateau stretching to the west of the volcanic cones, from the northwestern town of La Cruz to the southeastern town of Bagaces.

Orosi-Cacao Volcanic Complex

Dominating the south of Lake Nicaragua, the first volcanic cone emerging is that of the **Orosi-Cacao** ensemble. These are Upper Pleistocene stratovolcanoes composed of several cones and craters, including **Orosi volcano** (4724 ft), **Orosilito volcano** (3937 ft), **El Pedregal volcano** (3608 ft), and **Cacao volcano** (5442 ft). Seen from the north Orosi, the volcanic cone presents itself as a perfect, pointed cone coated on top of lush tropical vegetation, which shows its relative youth because tropical erosion has not fitted it with deep canyons. Craters of this set are runaway with open mouth toward the southwest, an effect of eruptions coupled with trade winds that blow constantly from the northeast. An explosive caldera structure separates the craters of Orosi and Orosilito. Northern flank cones, however, are furrowed by deep ravines covered by thick tropical vegetation. In contact with the volcanic complex is the northern town of Santa Cecilia; an ignimbrite plateau extends where a shape of flat-bottomed canyons and fluvial erosion terraces are modeled in the ignimbrite rock. From Santa Cecilia, the plateau is covered by lahars, creating a shape of a topographic plateau of multi-convex reliefs. Farther south, Cacao's volcano cone rises, attached to the structure of Orosis's complex. The southwestern flank, however, is characterized by two open and very eroded craters observing the ignimbrite plateau.

To the middle flank, a third recent crater is imposed on the precedents, where an important lahar flow of pyroclastic rocks emerged with a large amount of pumice, forming a big volcanic fan ending in contact with the structural plateau. At the foot of Cacao's eastern flank cone appears a vast circular depression that is without a doubt a collapsed caldera, partly eroded by the **Pizote River canyon,** also known as Niño River. A northeast-southwest tectonic fault limits the first volcanic complex with the Rincon de la Vieja volcanic structure (Figure 7.3).

Rincon de la Vieja volcanic complex

The Rincon de la Vieja volcanic complex is located between the Orosi-Caco and **Miravalles volcanic systems**. It is a stratovolcano complex dating from the Middle to Upper Pleistocene (Boudon et al., 1995). Nine craters are aligned on the elongated top of this massif. Between them are **Santa Maria crater** (6286 ft), which houses a pluvial lake, and **Rincon de la Vieja,** the main active crater of this massif, whose south wall reaches 5950 ft above sea level. Rincon de la Vieja presents great volcanic activity, of which the most recent eruptions date back to 1966-1970, 1991-1992, and 2011, with intense fumarole activity. In the northwest emerges **Von Seebach crater,** rising 2936 ft above sea level. The southeastern crater alignment is also characterized by the presence of a large explosive caldera at 5249 ft altitude, eroded and open toward the northeast, housing two craters. The edge of the caldera has another very eroded crater known as **Marmo crater,** reaching to 5380 ft above sea level. Though it is regular, the western slope presents a great erosion circus in its sector, where rivers have carved deep canyons (Figure 7.4).

In the base, dacitic to ryodacitic Lower Pliocene domes emerge (Bellon and Tournon, 1978): **Fortuna dome (**1571 ft), **San Roque dome** (1771 ft), **Cañas Dulces dome** (2148 ft), **Gongora dome** (2519 ft), and **San Vicente dome** (1968 ft) (Bergoeing, 2007).

FIGURE 7.3

Guanacaste volcanic range: Orosi, Cacao, and Rincon de La Vieja cones, with Santa Rosa's ignimbrite plateau in the forefront. *Photo courtesy of Gilbert Vargas, 2010.*

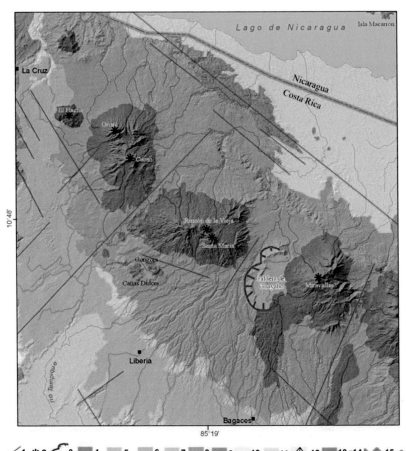

FIGURE 7.4

Guanacaste volcanic range geomorphology: 1—Tectonic faults; 2—Craters and volcanic cones; 3—Collapsed calderas; 4—Lahar fields; 5—Quaternary altered volcanic lava flows; 6—Tertiary sedimentary hills (Paleocene-Eocene); 7—Ignimbrite cone-glacis of Guayabo collapsed caldera; 8—Middle Pleistocene cones and craters; 9—Pliocene dacitic domes of Cañas Dulces; 10—Holocene sediments; 11—Ignimbrite plateau; 12—Alluvial cones; 13—Pliocene volcanoes; 14—Towns; 15—International border; 16—Rivers. *J.P. Bergoeing, 2012.*

Also, it is worth mentioning the thermal springs and mud sources associated with Rincon de la Vieja's volcanism at the Coyol site, bordered by the Colorado and Jaramillo Rivers. The contact between the volcanic cone and the western plateau is the result of mainly successive spills of ignimbrite and lahar. Eastern slopes, which are more regular, are covered by dense natural vegetation, ending in a volcanic structural plateau inscribed in a former collapsed caldera. This is probably not recorded in geological charts because of the permanent cloud cover of this sector, but it becomes evident when studying satellite radar imagery. The surface material of the past 300,000 years of this volcanic massif consists of cinder, lahar, and pyroclast deposits of andesitic composition (Carr et al., 1986; Chiesa et al., 1992). The historical eruptive activity of this complex dates back to 1765 and has been characterized mainly by steam and ash columns. According to Boudon et al. (1995), Rincon de la Vieja is able to produce eruptions with lahar spills (mainly toward the north), acid rain, volcanic avalanches, and ash falls.

Miravalles Complex

Miravalles summit volcano reaches 6653 ft above sea level, and its crater has an approximately 656-yd diameter. It has six volcanic sources aligned northwest-southeast. This stratovolcano was rebuilt on several occasions during the Quaternary. At its northwestern foot stretches a vast topographic plateau that corresponds to the bottom of the **Guayabo collapsed caldera,** dating from the Lower Pleistocene (between 1.6 and 0.6 million years B.P.). This caldera produced strong emissions of ignimbrites (burning clouds) and finally the collapse or subsidence of the magmatic chamber of the former Guayabo volcano (Gillot et al., 1994). The remaining rim is represented by **La Montañosa** hills. The caldera occupies an area of 77.220 mi². Northeast of this place, an even greater collapsed paleocaldera dominates the base of this colossus, separating it from Rincon de la Vieja massif, as evidenced by satellite radar images. The current Miravalles cone has been built on a vast collapsed caldera that is shared with southern **Tenorio volcano,** whose evidence is once again demonstrated by satellite radar images.

Miravalles' southwest flank is bordered by **Espiritu Santo** and **Gota de Agua** hills, which correspond to Lower and Middle Pleistocene eruptive cones of the former Miravalles. They are separated from the Miravalles main cone by a west-southwest-east-northeast tectonic fault passing between **Tenorio** and **Montezuma** volcanoes, disappearing in the northern plain. Remains of less important calderas exist in the Miravalles northeastern flank. Miravalles volcano is separated from Tenorio-Montezuma's volcanic complex by a southwest-northeast tectonic depression. Miravalles volcano is currently active with fumaroles, solfataras, and thermal springs. Holocene andesitic lava flows cover the northwest flank, and geothermal activity manifested since 1964 is today operated by the Costa Rican Institute of Electricity (ICE) (Figure 7.5).

Tenorio-Montezuma Volcanic Complex

Tenorio and Montezuma's cones are the southernmost volcanic complex of Guanacaste's range, if we take into account the gap between Guanacaste's range and the **Arenal** volcanic complex. Tenorio and Montezuma are two stratovolcanoes of the Middle to Upper Pliocene, composed of two main

FIGURE 7.5
Miravalles stratovolcano in the Guanacaste mountain range. *Photo courtesy of Gilbert Vargas, 2008.*

cones: Tenorio (1 mi 335.36 yd) and its twin Montezuma (5971 ft), separated by a southwest-northeast tectonic fault. Among them is equally drawn a small collapsed caldera of a third volcanic edifice that was beheaded, and whose remnants are a series of similar altitude cones composing a structural plateau. On the ground, the hills present a multi-convex morphological shape, composed of pyroclastic materials. Between the plateaus, one can glimpse depressions that could be remains of small explosive craters. Tenorio-Montezuma's volcanic complex would be related to the Miravalles modern building because all are circumscribed in a peripheral caldera that could date back to the Middle Pleistocene, and whose evidence can be seen in satellite radar images. Like Miravalles, the Montezuma volcano cone is separated in the south by a southwest-northeast tectonic fault of another volcanic structure of the Upper Pleistocene. It is quite eroded, but one can still glimpse the original morphology. Tenorio volcano dominates the vast tectonic depression of **Arenal Lake** from the northwest, which is a set of low volcanic mountain ranges of similar altitudes, dating from the end of Pliocene to the beginning of the Pleistocene, resting on Tertiary sedimentary series (Venado Formation). In the northern sector of Tenorio-Montezuma's set, a series of rivers and waterfalls associated with volcanic hydrothermal activity gives particular staining to **Celeste River** by the sulfuric input of its waters (Figure 7.6).

Los Perdidos-Chato-Arenal Modern Volcanic Complex
Los Perdidos, Chato, and Arenal volcanoes are recent. Built during the Upper Pleistocene and Holocene, they are close to the south Guanacaste volcanic range and serve as a transition to a new unit in the Central volcanic range.

FIGURE 7.6
Tenorio volcano with its totally destroyed peak from volcanic eruptions during the Holocene. *Photo courtesy of Francisco Solano, 2010.*

Arenal volcano, 5436 ft above sea level, is a modern cone built on the oldest explosion caldera. It is estimated that Arenal has only been in existence for 40,000 years, making it the youngest volcano in Costa Rica. Historical activity is not known, but since 1937, it has shown violent activity. It is a stratovolcano composed of an alternation of ashes, slag, lapilli and lava blocks, and deposits of *nuees ardentes*. Volcanic products range from basalts to dacites (Borgia et al., 1988). Since the 1968 eruption, it has manifested as a dangerous volcano, emitting pyroclastic flows. Its activity has not ceased, which has attracted a large number of tourists, mostly because of the spectacle of its unpredictable activity and its hot springs. For this reason, the construction of hotel facilities at the same foot of the volcano—in some cases 1.5 mi in a straight line from the crater—chronicles imminent disaster. Though the hotels have been legally authorized by the proper authorities, conditions are in place for a human catastrophe. In the event of a strong pyroclastic eruption, many hotels would be buried by the red-hot avalanche. The surrounding nonradial road is completely inadequate for a quick evacuation (Figure 7.7).

San Lorenzo Collapsed Caldera
This description would be incomplete without mentioning the San Lorenzo collapsed caldera. Indeed, such a large caldera is only perceived through satellite images, and probably fits in the Lower Pleistocene. This caldera marks the transition between the Central and Guanacaste volcanic ranges.

Central Volcanic Mountain Range
The Central volcanic mountain range is the result of the construction of five large volcanic edifices, built gradually during the Quaternary. They are the **Platanar-Porvenir, Poas, Barva, Irazu,** and **Turrialba** complex. They are active volcanoes and each has its own structure and geomorphology, with adventitious cones, *planezes*, lava flows, lahars, and thermal and gaseous emissions. Current craters are aligned following a northwest-southeast tectonic direction.

FIGURE 7.7
Arenal volcano in slow effusive activity. *Photo by J.P. Bergoeing, 2010.*

Central Valley is part of the Central volcanic mountain range and an extension of the same. It is a vast structural plateau in the western sector, bounded by Grande River and carved by the **Virilla River** canyon. **Ochomogo** pass separates Central Valley in two main sectors: western (including the cities of **San Jose, Heredia,** and **Alajuela)**, and eastern (including the cities of **Cartago** and **Turrialba)**. The Central plateau occupies a tectonic graben position, bordered by the horsts of the Central volcanic range and the **Talamanca** mountain range, the latter ending there by staggered failures. The Central plateau is part of the Quaternary volcanic basement, though its first manifestations occurred during the Upper Pliocene and date back to the Lower Pleistocene. Its base consists of a powerful series of basaltic lava flows. Its stratigraphy evolves into series of tuffs clearly appearing in the Virilla River canyon, covered by important ignimbrite deposits with an output of 114-328 ft thickness, divided into two units (Perez, 2000). This shows increasingly important acidification in the eruptive events of the Central volcanic range. Finally, the Central plateau is covered by numerous, important, and visible lahar deposits coming from the southern flank of the Central volcanic range, constituting hill cords perpendicular to Virilla River. The eastern sector, where the cities of Cartago and Turrialba stand, is constituted by a smaller plateau, generally of the same nature as western plateaus such as **Paraiso** and **Juan Viñas**. From Turrialba to Laguna de Bonilla, **Reventazon River** notches Eocene-Miocene marine sedimentary formations on its left bank, on which rest the oldest volcanic lava flows and lahars (Tournon and Alvarado, 1995). Central range stratovolcano cones are chronologically more modern than the Central plateau, sitting atop an understood margin between the Middle and Upper Pleistocene. Current craters are mostly Holocene.

Platanar-Porvenir Volcanic Complex

The **Platanar volcano** is located northwest of the Central volcanic range. It consists of a composed stratovolcano that can be dated tentatively from the Upper Pleistocene, constituted by the modern cone of Platanar and the **Porvenir volcano** cone a little farther south, separated by a distance of 1.5 mi. The volcanic complex covers a surface of more than 43,630 mi². Platanar's summit rises 7162 ft above sea level, while Porvenir's reaches 7437 ft. On Platanar's northern flank, overlooking the great north plain, the remnants of **Palmera's collapsed caldera** rise (Tournon, 1984), probably from the Lower Pleistocene, stuffed by lahar deposits in large part, forming large alluvial fans coming from Platanar's northern flank. This filling can be reasonably considered a constant stream, which stuffed the collapsed volcanic depression during the entire Quaternary period. Farther east, nine Upper Pleistocene volcanic cones appear aligned north-south, known as **Aguas Zarcas volcanism**. They reach elevations no higher than 368 ft. in altitude, but they have well-preserved cones, and some even have their craters. They are the manifestation of the most recent magmatic ascension of the sector, consisting of gabbros and dolerites, which correspond to a calco-alkaline Holocene volcanism. Some are Strombolian cones composed by lapilli, volcanic bombs, and alkali basalts (Tournon and Alvarado, 1995) (Figure 7.8).

Platanar and Porvenir volcanoes are modern constructions (Upper Pleistocene) rebuilt on older structures, where noteworthy. To the east, Chocosuela caldera forms a deep canyon, reaching 3280 ft deep. At the bottom runs the Aguas Zarcas River (Alvarado et al., 2006). If we consider that this whole structure rests on a wider structure, going from Toro River on the east side to

FIGURE 7.8
Well-preserved volcanic cone and crater of Aguas Zarcas. *Photo by J.P. Bergoeing, 2010.*

San Lorenzo River on the west side, then the area takes another dimension and the Platanar-Povenir complex fits within the framework of the world's largest calderas, because the diameter of the whole structure would give it a surface of approximately 386 mi².

Poas Volcano

Poas Volcano is a volcanic group of the Central Mountain Range, rising to 8884 ft in altitude. It consists of a main crater that is more than 1 mi 1520.8 yd in diameter, containing a sulfuric acid lake with a pH1 degree of acidity and temperatures of 185 °F. It also emits sulfuric gas. A very close second crater, which is by now inactive, houses a pluvial lake known as Botos lagoon. The whole area is surrounded by the dense tropical vegetation of the mountains. The Poas is also part of a tectonic system through which several magmatic issuers emerge in a north-south line. In addition, a few kilometers north of the Poas stands the inactive **Congo volcano** at 6607 ft, whose crater is open to the northwest. Following tectonic alignment, further to the north stands **Hule caldera** (Bergoeing, 1977), a volcanic depression that houses three small lagoons and in whose center stands a post-collapse volcanic cone. Finally, the whole group ends up farther to the north because of the **Laguna of Rio Cuarto gas maar**. In the plain of the north, there is an explosive crater depression, covered with fluvial waters. This gas maar is 1181 ft above sea level. Its walls are basaltic-andesitic, measuring about 0.12741 mi² at the surface, with a depth of 216 ft (Alvarado, 2000).

Poas lavas are complex. The southern flank presents afiric andesite lava, a rarity in the calco-alkaline series, which can be acidic or basic. The volcanic lava ranges from basaltic to dacitic. **Congo volcanic cone** lava and **Laguna de Hule caldera** lava are mostly basaltic-andesitic (Tournon, 1983). Since 1828, Poas volcano has had 39 known eruptive episodes. During the eruption of January 25, 1910, a steam column was expelled, detaching 1,410,958,592 lb of ash. From 1952 to 1954, Poas volcano presented a new cinder eruption cycle with slag emissions. In 1987, a phreatic explosion modified the lake level, and in 1989 the lake disappeared temporarily, exposing sulfur hotbeds. The crater was submerged again in 1990. In 2006, Poas again manifested phreatic explosions and fumaroles (Figure 7.9).

The rivers running to the north from this point have deeply eroded volcanic slopes, forming fluvial canyons, including the Sarapiqui River, an important tributary of the San Juan. The lava flows have allowed the creation of choppy modeling, where cascades and waterfalls are abundant. At the foot of the cone of the Poas, a transition with the northern flood plain is produced by powerful coalescing laharic fans of the lower Pleistocene (Bergoeing, 2007). Finally, the Poas massif is bordered to the west by the Toro River, which forms part of the Platanar-Porvenir collapsed caldera system.

FIGURE 7.9
Principal sulfuric acid crater of the lake of the Poas volcano. *Aerial photo by J.P. Bergoeing, 2012.*

Barva Volcano

Barva Volcano is separated from the Irazu-Turrialba volcanic complex by the crossing pass of La Palma and Zurqui's relict volcanism. Barva Volcano dominates the city of San Jose from its altitude of 9534 ft. It has about twelve eruptive outbreaks on its top and adventitious cones on its flanks. Three volcanic domes crown its summit, with Barva's main crater occupied by a pluvial lake at an altitude of 8228 ft above sea level. Laguna Danta is another northern pluvial lake crater at a similar altitude. It has had no known historical activity, although there are residual emissions of CO_2 (solfataras) and hot springs. Barva's lava is mostly comprised of granulites of andesite and basalt, with some olivine and dacite (Alvarado, 2000). One of southern Barva's slopes is the enormous Los Angeles basaltic-andesitic lava flow, contemporary with the Cervantes lava flow of Irazu Volcano, except the former spill covers a much wider extension, reaching the city of Barva. The lava flow originated in Cerro Redondo dome; therefore, this is one of the adventitious domes of Barva Volcano (Figure 7.10).

The western flank is dominated by the structure of the **Guariri volcano** (Protti, 1986), an explosive Holocene volcano that has emitted andesitic and pyroclastic lava flows that have reached Santo Domingo and El Roble. The crater, which is open to the west, is partly destroyed by erosion. To the northeast, Barva presents a collapsed caldera structure of large dimensions, but heavily affected by erosion and completely covered by tropical mountain vegetation. Despite not having a historical record of volcanic activity, Barva Volcano is dangerous because there are traces of two Plinian eruptions and ignimbrite spills, showing explosive acid activity (Paniagua, 1984). The old-

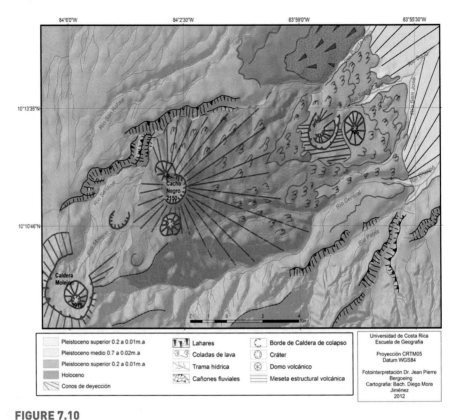

FIGURE 7.10
Geomorphology of Molejon collapsed caldera, Cacho Negro volcano, and Santa Clara caldera. *Based on satellite radar image by J.P. Bergoeing, 2012.*

est known Holocene eruption dates back to 6050 ± 1000 years B.P. (Simkin and Siebert, 1994), but the andesitic-dacitic tuffs exposed in Tiribi River date back 332,000 years, coming from Barva's caldera (Smithsonian National Museum of Natural History, 2007). Barva's northern slopes are also characterized by a series of calderas and domes, and Cacho Negro Volcano. Despite being a Holocene cone, it is very upset by erosion. From its bottom, Puerto Viejo River surges. **Cacho Negro stratovolcano** rises 7381 ft above sea level and is located northeast of Barva's summit. It is a fully open northwest crater, and on its flanks it is possible to observe recent (Holocene) radial lava flows and two adventitious cones (Alvarado et al., 1980). Lavas are predominantly andesitic and basaltic, but there is no known historical activity. The northern plain border with this volcanic complex is marked by **Santa Clara's collapsed caldera,** with two post-collapse cone inserts inside the caldera (Bergoeing, 2012a,b). Northern Barva slopes can be divided into three sectors: the western sector, composed of the Guariri volcano cone; the eastern sector, beginning with Zurqui hills and culminating with Cacho Negro Volcano; and a central sector, which is a massif with very regular slopes descending gradually to the northern plain, but interrupted by numerous rivers that fit it vertically. This is due to the volcanic material, which in this part of the northern slope is made from cinder deposits interspersed with lava flows and tuffs. It is perhaps in the northwestern part of the Barva Volcano slopes where you can better observe the mega-landslide phenomena associated with second and third landslide events, the product of a very thick altered mantle, formed by old clay material from the decomposition of volcanic deposits. This deeply altered material, which has been overstretched by water of the tropical climatic conditions on this flank and favored by steep slopes, has followed, causing mass landslide phenomena and subsequently second landslides over the already removed masse material (Figure 7.11).

Zurqui Hills

Located between Irazu and Barva volcanoes, the Zurqui hills are a volcanic group composed of **Chompipe** (7411 ft), **Turu** (7017 ft), **Caricias** (6889 ft), **Honduras** (6715 ft), **Tres Marias** (5659 ft), and **Achiotillal** (6174 ft) volcanic cones. They are located in the most depressed part of the Central Mountain Range and constitute *Paso del Desengaño* (disappointment pass), where humid Caribbean flows easily penetrate the Central Valley, which is why this volcanic ensemble is significantly eroded despite its relative youth. An andesitic radiometric date gave it an age of 5 million years (Bellon and Tournon, 1978). This geological survey shows that the Zurqui hills are significant sources of basaltic-andesitic lava flow emissions, tuffs, and breccias, as well as ignimbrites and pyroclastic material, which are exposed in Braulio Carrillo highway and Zurqui Tunnel. The hills are covered by a significant thickness of altered ash.

FIGURE 7.11
Main open crater of Cacho Negro Volcano. *Aerial photo by J.P. Bergoeing, 2012.*

Irazu Volcano

Irazu stratovolcano has the highest active crater of Costa Rica, with 11,259 ft of altitude. Situated north of the city of Cartago, it is also composed of multiple Holocene craters. It has some of the oldest cones, aligned northwest-southeast, ending on the destroyed **Las Nubes volcano** crater (Bergoeing, 1979). Irazu Volcano has historical activity: the colonial chronicles mention eruptions from the eighteenth century, with cycles of 40 to 60 years. Since 1723, it has had about 23 eruptions. The 1963 eruption had a lahar flow that buried the urban sector of Taras in the city of Cartago, after heavy rains associated with the volcanic eruption. San Jose and most of the Central Highlands of Costa Rica were then coated with ashes. Tournon (1983) identifies basaltic lava flows in the top as well as basic andesites in the Cervantes and Juan Viñas lava flows. Cartago's plateau is composed of ignimbrites interspersed among andesites (Figure 7.12).

Irazu's southern slope is divided into two important morphological sectors. The upper area is formed by an embossed, wavy, multi-convex landscape, with juts from the crags of recent lava flows of the volcano covering the southern flank to the town of Cot. This sector is characterized by deep thickness of ashes and highly altered ancient andesitic-basaltic lava.

The second sector extends from Cot to the volcanic plateau, where the city of Cartago stands on a series of laharic coalescent cones. This sector, north of Cartago, is comprised particularly of lahars, highlighting two specific generations: one very altered, ancient, ochre bottom made up of lapilli and pumice; and a second, more recent generation that is higher and has healthy, rolled, chaotic materials (Bergoeing, 2007). The southern slope is also characterized

FIGURE 7.12
Very eroded northern slopes of Irazu stratovolcano of the Central Volcanic Range under heavy tropical rains. *Aerial photo by J.P. Bergoeing, 2012.*

by the big Cervantes basaltic-andesitic canoe-type lava flow. It has been dated 23,000 years B.P. by the $^{238}Th/^{232}Th$—$^{238}U^{232}Th$ (Allegre and Condomines, 1976) method. But other radiocarbon (^{14}C) dating rejuvenates it to 20,000 years B.P. (Murata et al., 1966). The Strombolian eruption that caused the Cervantes lava flow (Tournon, 1984) came from the **Pasqui** volcanic cone and resulted in naturally damming the waters of the Reventazon River in the area of Cachi. Four lake terraces in this sector were built by the damming.

The northern slopes are covered with dense, primarily tropical mountain vegetation, which is very different from the southern slopes. The upper part of Irazu's cone is characterized by "Lavakas," a Malagasy term designating an erosion mega-circus. In fact, the northern slopes are modeled by deep scarps of 900-1600 ft of free fall, on whose walls are emissions of brimstone because of their proximity to the volcanic chimney, a product of the erosive wear of rainwater. Lavakas are the first stage of a greater phenomenon found in the west on Poas and Barva volcano slopes, corresponding to mega-landslides produced during the Middle Pleistocene, inside of which are minor landslides that continue today. It is partly the forest cover that stopped the phenomenon, but it could be reactivated with the human colonization of the sector. Some major rivers like **Sucio River** undermine important river canyons that can reach 3000 ft deep in some places. This is possible because this stratovolcano material is very heterogeneous, comprising cinders, tuffs, and lava flows.

Volcanologist Phillip Ruprecht of Columbia University, New York, studied Irazu's 1963-1965 eruption lavas and classified the volcano as a "supervolcano" because its magma chamber is directly connected with the mantle; therefore, he dubbed it *"the highway of hell."* He arrived at this conclusion after checking the presence of trace amounts of nickel in olivine in lavas that should have melted before being ejected by the eruption, which is proof of the rapid magma ascent from the mantle (Ruprecht and Terry, 2013).

Turrialba Volcano

Turrialba Volcano is the easternmost stratovolcano of the Central volcanic range of Costa Rica. It rises 10,958 ft above sea level and at its summit presents a series of four craters, of which the three main ones present activity with fumaroles and sulfur gas fumes. The volcano erupted several times in the nineteenth century (1853, 1855, 1864-1866), with pyroclastic and intense fumarole emissions. Its ashes arrived at Nicaragua's Port of Corinto. The current top cone, like the others in the Central volcanic range, were built during the Holocene. It presents lava ranging from basalts to dacites. It is estimated that in the past 3500 years, there were at least five large eruptions with lava flow emissions. In 2007, its activity increased dangerously, with sulfur gas emissions affecting the local population. The top of Turrialba's slopes presents a series of "planezes" or tilted structural plateaus, formed by considerable lava thickness descending gently in the shaped skirts of the volcano. They are present in the south, east, and north. **Dos Novillos** hill in the eastern slope corresponds to a volcanic cone, the easternmost one of the Central Mountain Range. The oldest of Turrialba's basaltic-andesitic lavas goes back to 2.15 M.y. (end of Pliocene). One of the most interesting of Turrialba's lava emissions is **Peralta's lava flow** in the southeastern slope, which draws a curved ending plateau, reaching an altitude of 1968 ft. During the Upper Pleistocene it damned the Reventazon River runway, creating a temporary lake. Today, **Bonilla's lakes** are the only witness of this event (Figure 7.13).

From that point and heading toward the north, the lower part of Turrialba's cone is composed of powerful coalescent alluvial fans consisting of thick lahars from the volcanic activity of this massif. Turrialba's lava flows are similar to those of Irazu Volcano, although the basalts do not abound, and they are mostly composed of basic andesites (Tournon, 1983). Coalescing volcanic cones descending from Turrialba Volcano evolve gradually to alluvial fans, and finally to a glacis that gradually merges with the great north plain (Nicaragua's graben). Interestingly, also of the Upper Pleistocene are lava flows coming from Turrialba's cone to the city of Guapiles. In fact, a few miles south of the

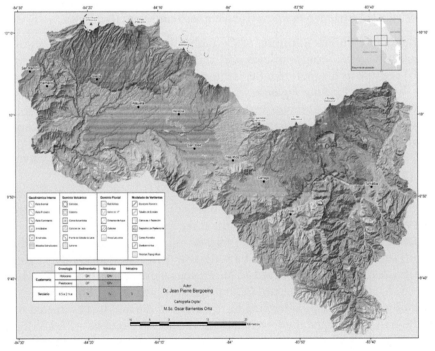

FIGURE 7.13
Costa Rica's Central Valley geomorphology. *J.P. Bergoeing, 2012.*

FIGURE 7.14
Turrialba Volcano with its four craters. *Aerial photo by J.P. Bergoeing, 2012.*

city stand three imposing andesitic lava flows resting on old laharic flows that are part of the aforementioned coalescing cones, where underground water springs surface, watering the city of Guapiles. These could also water supply the whole Central Valley. South of the crater is the city of Turrialba, seated on a major tectonic accident (north-northwest-south-southeast failure) covered with important laharic flows. The southern slopes constitute the northern flank of the eastern Central Valley. In it sits an ancient explosion caldera structure whose remnants extend from Coliblanco to Juan Viñas (Figure 7.14).

TEMPISQUE RIVER'S TECTONIC DEPRESSION

Tempisque River's tectonic depression covers the north of Nicoya Peninsula. It is a tectonic subsidence zone pinched between the Guanacaste volcanic range and Nicoya Peninsula, both in horst position. The tectonic depression is oriented northwest-southeast and is traversed by Tempisque River, which flows into it, leaving the ignimbrite plateau through the river gorge. River sediments feed the plain, which is how it has been partly sealing this depression. However, positive tectonics affecting this entire sector have raised the vast area, which was submerged in the sea during the Late Miocene and formed a broad interior gulf. Today, the Gulf of Nicoya is a pale remnant of it. Near Tempisque River estuary rise a series of isolated hills ("mogotes"), which are a testimony of old coral reefs formed during the Miocene. The positive effect of plate tectonics is evidenced by isolated hills known as **"Cerros de Barra Honda."** Inside are karstic caverns, a result of the infiltration of tropical rains in the last remnants of the coral limestone. Today, the Tempisque River depression is a very fertile and vast plain, exploited by agricultural activities. In the western Filadelfia village sector stands a series of isolated hills whose structures reveal the last fluviomarine

FIGURE 7.15
Low course of Tempisque River near its sea mouth. Mangroves and swamps protecting a wild fauna.
Photo by J.P. Bergoeing, 2009.

episode formed by rolled pebbles and covered on the top by Quaternary ignimbrite deposits, an acidic witness of the maximum extent reached by the volcanic manifestations of Guanacaste range volcanism (Figure 7.15).

Karstic Formations

Karst is an underground formation created by limestone dissolution plateaus. This concept was introduced in 1893 by Serbian geographer Jovan Cvijic, a specialist in geomorphology. Costa Rica and the other Central American countries are not excluded from this phenomenon.

In Costa Rica, on the right bank of Tempisque River, forming an estuary near its mouth and emptying in the Gulf of Nicoya, emerges a former reef relief known as Barra Honda hills. The hills have been uplifted by the Quaternary orogeny and sheltered inside vertical caves fed by tropical rainwaters, creating stalactites. Water springs appear at the hill base, forming stepped travertine by saturation of calcium magnesium carbonates (Figure 7.16).

Another sector where the phenomenon of karst dissolution is visible is in Venado, a few kilometers from Arenal Volcano. Contrary to Barra Honda, here we see a series of horizontal caverns carved by the water in the limestone of the Venado Formation, dating from the Miocene. The caverns are also inhabited by bats, whose excrement can be dangerous for human beings because it can cause respiratory problems. The entrance to the caves is made with difficulty, and the visitor can get lost easily. It is necessary to have headlights, as the darkness of the interior is total. Of course, stalactites and all the karst phenomena are visible inside (Figure 7.17).

FIGURE 7.16
Stalactites formed by dissolution of the limestone of Barra Honda (Damas cave) due to the infiltration of rainwater, which after a journey inside the caves appears as springs, loaded in calcium and magnesium carbonates, giving origin to travertine, where it stunts in stairs. *Photo courtesy of Gustavo Quesada, 2010.*

FIGURE 7.17
Venado Caves, Costa Rica. Stalactites hanging from the ceiling of the caves inhabited by bat and travertine steps. The cave system runs approximately 3 mi and is conditioned by the tectonic and underground rivers. *Photo courtesy of http://www.hotelarenalregina.com/es/arenal-volcano-tours/15-venado-caves.html.*

TERTIARY MOUNTAIN RANGES

Tilaran Mountain Range

The 31-mi-long Tilaran Mountain Range is a mountainous unit with a much smaller surface than the Talamanca or Central ranges. Its natural boundaries are **Arenal Lake** to the north; to the northeast, San Lorenzo River serves as the border with the Central volcanic range like Aguacate mountains to the south, in contact with Talamanca range. Tilaran extends south to Tarcoles River, if we consider Aguacate mountain (in genesis) forming an integral part of this unit, which it does because of its geological history and the nature of the rocks and their ages. We can also include the mountain range that lies to the south of Tarcoles River and Santiago de Puriscal. This sector ends with the Cretaceous-Tertiary isolated **Turrubares volcanic peak** (5761 ft) overlooking the Pacific Coast. Tilaran range is almost an exclusive product of Miocene-Pliocene volcanic activity, whose deposits have been regrouped under the name Aguacate Group of Formation (Madrigal, 1972), which includes a large number of effusive rocks. Most of the volcanic centers of this mountain range have disappeared or are unrecognizable due to erosion. However, some elements like domes, craters, necks, and calderas remain, such as **Palmares collapsed caldera** or **Pelon-Mondongo-Tinajita** cones and craters (Bergoeing and Brenes, 1978a,b), as well as ancient volcanic hot spots like Cerro Tilaran, Cerro Pelado-Delicias, Chopo, Los Perdidos, Poco Sol, Pelados-Herrera, Espiritu Santo (Mora, 1977), and Cerro Macho Chingo-Pelon (Alvarado et al., 1980). The aforementioned authors place these last volcanic emission centers on the map of "Plio-Pleistocene volcanism" (Figure 7.18).

This volcanic range follows the general direction of the country, whose axis is northwest-southeast. It has two well-defined aspects: the northeastern slopes (or San Carlos slopes), which are tributaries of San Juan River, pouring water into the northern plain following a sinuous path; and the southwestern slopes (or Pacific slopes), which are subdivided into three sectors:

> **Northern sector, between Cañas and Barranca River**
> **Southern sector, between Barranca and Tarcoles Rivers**
> **Turrubarres Hill sector**

Already in the western Central Valley, the Palmares sector is where we find the most spectacular collapsed caldera of this area. It is an ancient Pliocene volcanic building that collapsed at the beginning of the Quaternary, with the testimony of volcanic dykes in the rim. Post-collapse volcanism continued, however, giving rise to the **Espiritu Santo volcanic cone** (4438 ft above sea level). However, Palmares collapsed caldera history extends

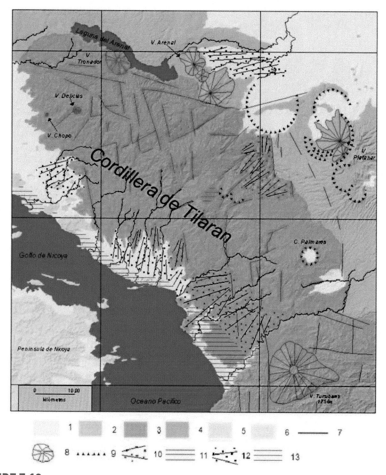

FIGURE 7.18

Tilaran Mountain Range geomorphology: 1—Quaternary sedimentary area; 2—Quaternary volcanism area; 3—Tilaran volcanic area (Pliocene); 4—Cretaceous-Tertiary volcanism; 5—Tertiary sedimentary area; 6—Cretaceous-Tertiary sedimentary area; 7—Tectonic faults; 8—Volcanic cones; 9—Calderas; 10—Lahar fans; 11—Structural plateaus; 12—Alluvial fans; 13—Floodplains. *J.P. Bergoeing, 2011.*

over the Lower Pleistocene because it housed a lake. Volcanic ash was deposited in its waters from the active volcanic cones in the vicinity (Espiritu Santo, Chayote volcanoes). Some ashes were locked in a small spherical clayed matrix, which was brown because of the rotation effect of lake waters, and deposited on the bottom. Before becoming completely full of sediment, the lake emptied eastward through the Grande River tectonic fault, which took advantage of the crushing rock area to dig a deep canyon (Figure 7.19).

FIGURE 7.19
Palmares collapsed caldera. *Photo by J.P. Bergoeing, 2013.*

Talamanca Mountain Range

The highest point of the country is in southern Costa Rica, with **Chirripo mountain** at 12,529 ft above sea level. Only part of the Talamanca range is volcanic (Miocene-Pliocene). The core is formed by intruded granodiorite, forming a batholith during the Miocene among Tertiary marine folded and faulted series. Little by little, the Talamanca mountain range has been hinting at its volcanic past, which is becoming clear as a result of intensifying research. In the Caribbean slopes, Tournon (1980) describes for the first time a series of explosion calderas along the Matama range. The observations of satellite radar imagery and land research identified a series of volcanic structures (Figure 7.20).

This did not escape the expert eye of Jean Tournon, a geologist at the University of Paris VI who recorded it in his geological map published in 1995 in Dieppe, France. The mentioned sector was punctuated by a series of basalt structures dated Pliocene, associated with intrusive, formed mostly by diorites and ledge monzonites as well as granite and gabbro. The indicated information deals with vast marine sedimentary formations composed by sandstone, shale, and conglomerates. Volcanic structures refer mostly to collapsed calderas of variable dimensions, situated around 3 mi in diameter.

Continuing south, in the coastal range of the Pacific watershed is old volcanic structure alignment from the Miocene to the Pliocene, going from **Mano de Tigre volcano**, through **Doboncragua** craters, to **China Kicha volcano** (Bergoeing et al., 1978a,b). From the Paso Real sector to China Kicha, on its border with the General River, the coastal range is characterized by a volcanic modeling that has been significantly altered by erosion. The sector is known as the Paso Real Formation, composed of volcanic conglomerates, with an age of 10 million years. Finally, Talamanca mountain range, at its southern end,

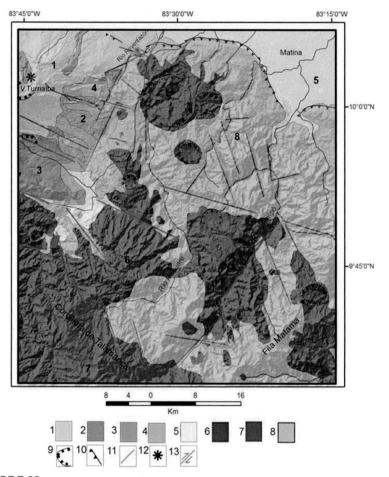

FIGURE 7.20

Caribbean watershed of the Talamanca mountain range, with volcanic structures that date back to the Pliocene formed in a marine environment before the orogeny: 1—Holocene volcanism; 2—Upper Pleistocene volcanism; 3 and 4—Lower Pleistocene volcanism; 5—Quaternary sedimentary plain; 6—Pliocene volcanism; 7—Miocene intrusive (granodiorites); 8—Tertiary sedimentary folded formations; 9—Calderas rim; 10—Reversed fault; 11—Normal fault; 12—Volcano crater; 13—Sliding fault. *Satellite radar image modified by J.P. Bergoeing, 2010. D:CR_Radar/CR_Hillshades.*

conceals a series of small volcanic cones that are the extension of the volcanic system previously described between China Kicha and Mano de Tigre volcanoes. Talamanca's Pacific area is characterized by dissymmetrical mountainous slopes, with a steep slope that falls to the Pacific and more gentle slopes that are directed to the Caribbean. It is an eminently volcanic area where **Cerro Fabrega caldera** stands open to the west, as well as **Frantzius** (7001 ft),

Pittier (9330 ft), and **Gemelos** (8864 ft) volcanic cones and other southwestern volcanic buildings. These recent volcanic cones could be parasitic cones of the Fabrega collapsed caldera. The volcanism that originated in the Upper Miocene lasted until the end of the Pliocene and likely reached the Lower Quaternary. It was formed of basaltic to andesitic rocks at the moment in which Talamanca's orogeny began. The oldest rocks in the sector are sedimentary; shale and conglomerates of the beginning of the Tertiary (Paleocene-Eocene) are folded, forming synclines and anticlines, and in the contact sector they are covered by volcanic deposits of the end of the Tertiary. **Irkibi volcano** could be accessed through Las Alturas; this cone is located east of Frantzius volcano following a magma extrusion alignment indicated previously. It is composed of several summits that leave us to assume other many craters where lava flowed. Mount **Chai** cone (6889 ft) is with Irkibi (7250 ft), and farther east is Cerro Bellavista (6719 ft), a volcanic package only fitted by the **Cotilo River**. At the foot of Chai hill abounds lava flows composed by dacites, according to Central American Geology School analysis. The western Bellavista River sector is a compact new volcanic set cut by the Cotilo River; the summit of the Fila Cedro range of 1.324 mi above sea level draws an old caldera. All of these volcanoes, limited in the south by the Coton River, are probably Pliocene-Pleistocene. Among them are deposits of granodiorite rocks that would indicate a Pliocene volcanic extrusion through the granodioritic Miocene Talamanca batholiths, and would beextended northwest with the volcanic complex **Mano de Tigre—Doboncragua,** dated as Pliocene (4-5 million years ago by K/Ar, Kessel, 1983 in Alvarado, 2000) (Figure 7.21).

In conclusion, we can say that Costa Rica is eminently a volcanic country, and its structural construction is due to the clash of the Cocos and Caribbean tectonic plates, which have already resulted in volcanism and orogeny that have lifted the whole country, rising it to a maximum altitude of 12,529 ft (Cerro Chirripo). This allowed the presence of Illinoian (Riss) and Wisconsinian (Wurm) glaciations in the Upper Pleistocene (Bergoeing, 2011a,b). Indeed, different field studies have determined the presence of huge erratic blocks in the General Valley that are the result of glaciers melting from the higher summits during interglaciation periods, as well as moraines, tarns, and U-shaped short valleys, which are strong evidence of ice parking suffered by high peaks during the Upper Pleistocene, when they exceeded 1 mi 1500 yd. elevation due to orogenesis (Figure 7.22).

THE PACIFIC COAST

Costa Rica's Pacific Coast is composed of four peninsulas: **Santa Elena, Nicoya, Quepos,** and **Osa.** Here we can find the oldest volcanic outcrops of the country, dating back to the Cretaceous. The Santa Elena Peninsula can be

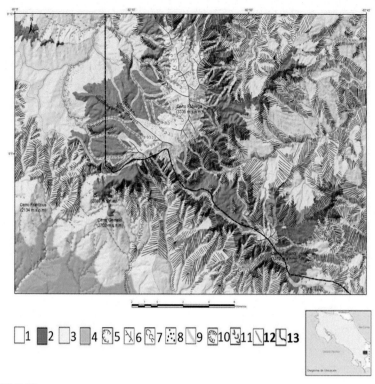

FIGURE 7.21

Maximum glacial extent in Fabrega's summit during Wisconsinian glaciations: 1—Illinoian and Wisconsinian paleoglacier parking; 2—Same age paleosnow parking; 3—Paleo *landes* area; 4—Paleosavannah; 5—Glacier circuses; 6—Edges of paleoglacier origin; 7—U-shaped valleys; 8—Moraines; 9—Paleoglacier valleys with moraines; 10—Slopes of erosion; 11—V-shaped valleys; 12—Watershed division lines; 13—International border of Costa Rica-Panama. *Geomorphology of J.P. Bergoeing, 2011.*

FIGURE 7.22

Chirripo tarns or glacier lakes surrounded by Wisconsinian moraines. *Photo by J.P. Bergoeing, 1982.*

defined as a seabed flake of green metamorphic rocks, where the main rock is a serpentinized peridotite. The Nicoya Peninsula, Osa Peninsula, and Quepos promontory are the oldest remnants of a predominantly basaltic insular volcanism of the Late Cretaceous. At the tip of the Nicoya Peninsula in Montezuma beach, a lumachelle conglomerate was lifted 22 ft above sea level by radiocarbon dating (^{14}C) (Giff/Yvette, France), showing a strong positive neotectonic marking the beginning of the Flandrian transgression 6.620 ± 150 years B.P. (J.P. Bergoeing, 2007).

The central Pacific Costa Rican coast is both a tectonic and climatic transition area. Indeed, in the southern Tarcoles River, the influence of Cocos and Caribbean tectonic plates associated with the Panama subplate and Nazca plate is evident. It is therefore an important seismic zone.

The Quepos promontory is the only notable accident that breaks the Flandrian coastal straightedge sandbanks of the central Pacific littoral. In Quepos, it is possible to see the base of the coastal Cretaceous volcanic rocks characterized by basalts (pillow lavas) and pelagic limestone, as well as shale. However, in both sides of this promontory, large coastal ridges and plains developed where the Parrita and Naranjo Rivers have built deltas, covered by mangroves belonging to the post-Flandrian transgression. But the country's largest mangrove is located at the Grande de Terraba River mouth delta. The main channel of this river has been running gradually northward due to neotectonic tilt. The Terraba River mouth is covered by a vast mangrove swamp built during the Flandrian regression 6000 years ago (Figure 7.23).

FIGURE 7.23
Terraba River delta, the biggest Costa Rican mangrove area in the Pacific littoral. *Aerial photo by J.P. Bergoeing, 2010.*

Osa Peninsula, which is less developed than Nicoya Peninsula, is mainly an uplifted plateau formed by sedimentary and volcanic rocks, resting on a Cretaceous basaltic basement. Osa Peninsula is tilted toward the northwest as well as toward Burica Point because of the effect of the clash of the Nazca plate with the Cocos plate. The result is that the greatest depths are found in Golfo Dulce's northwest sector, and not in its mouth (Figure 7.24).

Cocos Island in the Pacific

Cocos Island is located in the Pacific between 5° 29′52″ and 5° 33′50″ north latitude and 87° 01′44 to 87° 06′23″ west longitude. It belongs to Costa Rica and is situated 308 mi west of the Costa Rican side of Blanco Cape. The island is characterized by a humid tropical climate with 275 in. of continuous rainfall, and the average temperature hovers around 47 °F. An important network of water—a result of the humid climate—developed on the basaltic soils of the island. The Yglesias River is the most important, and drains the eroded cone of the Yglesias volcano. Rivers have created a V-shaped model with waterfalls flowing from the high cliff, a vivid result of coastal erosion.

The island was discovered in 1526 by Spanish sailor Juan Cabezas, and was visited throughout the centuries by sailors such as Wafer (1699), Duret (1720), Colnett (1788), Morell (1832), Coulter (1836), and Belcher (1836) (Foreign Office, 1919). It was also a penal colony between 1879 and 1881, and then a sporadic agricultural colony (1894-1905) under the orders of German August

FIGURE 7.24
Osa Peninsula's Cretaceous marine platforms crossed by numerous fissures. *Photo courtesy of Francisco Solano, 2010.*

Gissler, who abandoned it after the failure to settle. In reality, he unsuccessfully sought the Lima treasure hidden in the island. Today, the island is a National Park of Costa Rica (Figure 7.25).

Cocos Island is the original volcanic result of the hot spot of the Galapagos, and it is primarily formed by basaltic lavas. The Galapagos Islands and the aseismic ridges of Carnegie, Cocos, and Malpelo now form two separate and distinct tectonic plates. The island is the emerged result of the tectonic Cocos plate, which in the Pliocene collided with the Caribbean plate, giving origin to the Isthmus Central America. The seabed between Costa Rica and the island forms a volcanic ridge, where a series of submerged volcanic cones stand (Hey et al., 1977). The age of the island would be between 10 and 15 million years old, but according to Tournon dating (Saenz, 1981), the island is just 2 million years old, formed in the Pleistocene. The same age was found by Darymple Levy in 1970. The lava flows are formed of trachyte and basalts with olivine and a little hornblende. It can be said that the island has undergone two major volcanic episodes. The Yglesias volcano massif (1886 ft) has been the main center of emission and also the youngest one, and can be dated from the Middle Pleistocene (1.2 MA). It is shaped by two eroded craters and volcanic domes, where the highest points are Jesus Jimenez (1410 ft) and Pelon (1640 ft). To the east of this volcanic massif extends a structural plateau known as Llanos de Palo de Hierro, which rises 984 ft above sea level.

FIGURE 7.25
Cocos Island and Yglesias volcanic massif with a crater opening to the south. Cliffs formed by tubular basalts. The island is covered by a dense tropical rainforest. *Photo courtesy of Michel Montoya, 2004.*

East of the island, an earlier volcanic structure developed between 2 and 1.2 million years ago, running from the Lower to the Middle Pleistocene, characterized by three significantly eroded domes: Escorpion (918 ft) and Venado (787 ft) to the south, and a Chattam crater-shaped structure to the northeast. Weston Bay as well as Pajara Island could also be remnants of an old volcanic structure.

The island is also characterized by sharp cliffs that suffer the perpetual onslaught of the sea and that can reach more than 650 ft of free fall, sometimes ending in small detritus cones and containing many caves. Discontinuation of volcanic activity since the Upper Pleistocene permitted a strong erosion in land and in the coastal fringe, leaving a large, scattered number of small islets. The island is covered by a thick rainforest that partially prevents erosion of soil. Two major tectonic accidents (faults) severed the island from side to side, taking a northwest-southeast direction. While the first accident cut Massif Yglesias, the second installed Gema River. Similarly, there are other tectonic accidents with a northeast-southwest direction that cut Gema River and Escorpion dome obliquely. Here, the water frame is radial to volcanic systems (Figure 7.26).

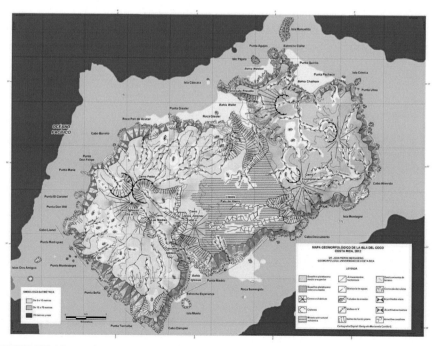

FIGURE 7.26
Cocos Island geomorphological map. *J.P. Bergoeing, 2012.*

THE CARIBBEAN COAST

The Costa Rican Caribbean coast is characterized north of the Port of Limon by a straight coastline, contrasting with large bays drawn south of this point. The straight coastline is formed by Flandrian parallel straightened sandbanks, aligned by the strong littoral drift. Inside, there is a series of navigable canals known as **Tortuguero canals,** dipped by dense tropical riparian vegetation, a silent witness of the recent surge of this coast. In fact, we can say that Limon's northern plain, down to 32 ft, was submerged during the Flandrian marine transgression 6500 years ago, and has since emerged with the help of positive neotectonics, as demonstrated by the 1991 Limon earthquake that raised the land 3-5 ft in some places (Figure 7.27).

In southern Limon, the coastline becomes narrower and forms an ancient reef fringe lifted by neotectonics, giving rise to a bordered coastal terrace evolving to an uplifted littoral tombolo where Punta Cahuita's resort stands and where the only living coral reef of this sector of the coastal strip exists. The southern Cahuita coral reef emerges from the coast and extends to be more developed in Bocas del Toro Peninsula in Panama.

Recent geomorphological studies in the northern sector of Costa Rica's Caribbean facade allow us to conclude that the contact between Lake Nicaragua and the Caribbean Seawas definitely closed during the Eemian period. In

FIGURE 7.27
Tortuguero's Flandrian coastal ridges, channels, and a small volcanic dome of the Tortuguero volcano (left). *Oblique aerial photo by J.P. Bergoeing, 2011.*

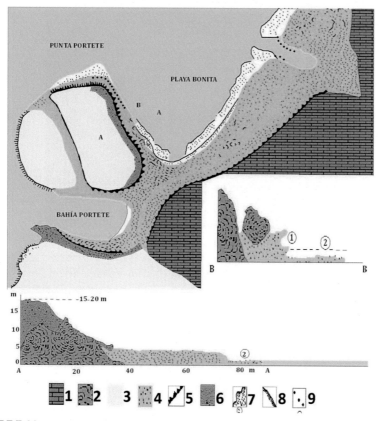

FIGURE 7.28

Coastal geomorphology of Portete-Playa Bonita beach sector in Limon: 1—Miocene strata; 2—Coral Portete formation; 3—Old sea level of 16-22 yd; 4—Puerto Viejo coral formation (Eemian); 5—Eemian cliff; 6—Red clay from Portete formation; 7—Corrosion marine platform on Puerto Viejo formation; 8—Today's active cliff; 9—Old Holocene sea level. *J.P. Bergoeing, 2013.*

effect, Caño Negro's sector constituted part of Nicaragua's paleolake until the Holocene, when it was sedimented by constant contributions of volcanic material transported by rivers from the Quaternary volcanic mountain ranges of Costa Rica (Figure 7.28).

Before coming into contact with the San Juan River margins, during the Flandrian transgression, the glacis (a natural extension of alluvial fans) flooded to the Caribbean. Later neotectonics raised the area, isolating this part of the country from the Caribbean Sea. Knowledge of the coastal relief changes during the Holocene interested not only the country in particular but also the international scientific community. During the Flandrian period (−6500 years), the

FIGURE 7.29
Coastal ridge of Puntarenas, Costa Rica, showing fine stalk that attaches it to the coast and the city that it has developed on the strip of sand that has no more than 9 ft of altitude. Behind the mangroves and in the foreground, the refraction of sea waves before reaching the coastline. *South Puerto Caldera, Google Earth 2006 satellite image.*

sea penetrated deeply into the coastal low sectors of Costa Rica and the rest of the world, marking their print in fossil marine deposits (such as Montezuma's lumachelles), coastal cords, and lagoons, becoming today's marshes and lakes in the process of drying. However, in many cases these testimonies are now in places that are more than 32 ft high, even though the world's Flandrian sea transgression did not exceed 13 ft high. Because of positive neotectonics, they are very active in Costa Rica due to the collision of the Cocos and Caribbean tectonic plates (Figure 7.29).

CHAPTER 8

Geomorphological Characteristics of Panama

STRUCTURAL GEOMORPHOLOGY

Panama's tectonic microplate is a product of the confrontation of the South American tectonic plate with the Caribbean plate and the Nazca and Cocos plates in the Pacific. Cocos plate subduction under the Caribbean plate began at the end of the Tertiary, provoking a generalized orogeny accompanied by strong volcanism. This determined the formation of a volcanic island arc between Costa Rica and Panama, originating the Talamanca-Tabasara mountain range and **Veraguas** and **Cocle** ignimbritic mountains, as well as the volcanic complex of **Valle de Anton**. It is a high seismic risk area. We can divide the country into two main sectors cut by the Panama Canal depression. To the west lies **Tabasara mountain range**, an extension of Costa Rica's Talamanca range, leaving two watersheds, as well as Bocas del Toro coral islands in the Caribbean and the Pacific plain of David, ending in the **Azuero** Peninsula before reaching the Panama Canal. The eastern sector of Panama is a vast depression surrounded by two emerging mountain ranges (Figure 8.1).

PANAMA'S VOLCANIC SYSTEM

Baru or Chiriqui volcano is the most important and active of the southern volcanic alignment of the Talamanca-Tabasara mountain range. This volcanic alignment extends into the Republic of Panama, inscribing about ten stratovolcanoes, starting in the northwest with **Cerro Fabrega caldera** and Tisingal cone, and ending in the southeast with **La Yeguada volcano**. It is also a highly tectonic sector with magmatic ascent, because the Caribbean, Cocos, and Nazca tectonic plates are confronted with Panama's microplate. This would explain the magma migration from west to east of some volcanic centers (Figure 8.2).

Baru or Chiriqui Volcano

Baru volcano is located to the west of Panama near the border with Costa Rica and constitutes its highest point, with 11,397 ft of altitude. Baru volcano seems to form part of the Talamanca Mountain Range; however, it is the product of

121

Geomorphology of Central America. http://dx.doi.org/10.1016/B978-0-12-803159-9.00008-X

FIGURE 8.1
Playon Chico island, forming part of the San Blas Island system on the eastern Caribbean coast of Panama inhabited by the Kuna natives. *Photo courtesy of www.choco.story.be.*

FIGURE 8.2
San Cristobal basaltic volcano on David's airport. *Photo by J.P. Bergoeing, 2009.*

the game of the tectonic Panama microplate activated by the underlying Nazca and Caribbean plates. The location of Baru volcano is explained by its being in one of the most seismic zones of the region—where the Caribbean, Cocos, and Nazca tectonic plates collide. However, some geologists have suggested that volcanism is inactive in the western region of Panama because the Cocos Ridge voids the subduction and magma ascents (Malfait and Dinkelman, 1972). However, recent studies supported by radiometric data show that volcanism was active during the Quaternary period, the Holocene, and the current period; the last catastrophic eruption dates from 740 ± 150 years B.P. Studies by IRHE-BID-OLADE in 1985 distinguish two eruptive cycles: the first is predominantly of lava flows, and the second is of phreatomagmatic explosive character, with pyroclastic flows (Camacho, 2007). This stratovolcano has

FIGURE 8.3
Baru volcano seen from David's littoral plain. *Photo by J.P. Bergoeing, 2009.*

no historical volcanic activity; however, it is a young volcano, because there is evidence of eruptions around 500 A.D. as well as minor eruptions around the year 1550. The samples obtained by radiocarbon dating (^{14}C) in volcanic ash-buried archaeological remains indicate ages ranging from 60 B.C. to 1210 (Stewart, 1978; Linares, 1975). Lavas are predominantly andesitic and basaltic (Figure 8.3).

The strong activity of Baru volcano is demonstrated by ash deposits in the Costa Rican area of San Vito de Java, which have given rise to a topographic plateau. Radiometric data carried out by the Costa Rican Institute of Electricity (ICE) revealed an age of 2.6 Ma (Bergoeing, 2007). Google Earth 2007 satellite image analysis shows two concentric explosion calderas: one ancient and peripheral; one modern and open to the west, where the volcano has built the current cone. The volcano has four major craters and two upper domes. The north-east flank, occupied by the villages of Las Mirandas, Cerro Punta, Guadalupe, Chumbaga, and Bambito, is a product of a collapsed caldera depression. The Baru or Chiriqui volcano is part of the southern volcanic alignment of the Talamanca range, which extends into the Republic of Panama, making up about 10 stratovolcanoes, beginning with Tisingal's northeast crater and ending with La Yeguada volcano in the southwest, the latter dating back to the Middle Pleistocene 1.38 Ma (MacMillan et al., 2004). Baru is a Plio-Quaternary modern stratovolcano built on a collapsed caldera. The current cone is located on a previous structure, open to the west. At the foot of the cone stands an important recent laharic flow (historical) that covers a large area. The resulting geomorphological pattern has not deleted multiform volcanic structures normally affected by tropical erosion, allowing us to recognize the previous volcanic relicts. This demonstrates that these are recent (Middle Pleistocene to Holocene), because the damp tropical erosive environment tends to destroy the old structural forms relatively quickly. The volcanic-alluvial fans that carpet the terrain south of the Baru volcanic cone can be ordered in several generations, according to a sedimentary classification. At the foot of the volcano are

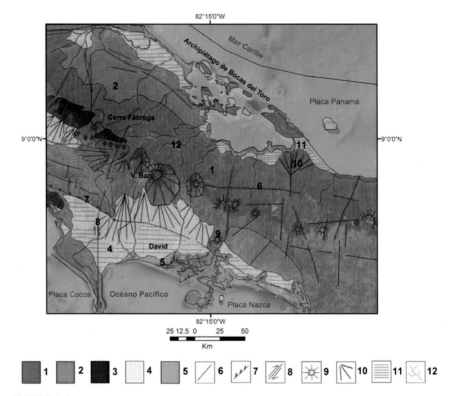

FIGURE 8.4

Structural geomorphology of the northwestern sector of Panama, bordered with Costa Rica: 1—Tertiary Quaternary volcanic area; 2—Folded Tertiary sedimentary area; 3—Tertiary intrusives; 4—Quaternary sedimentary area; 5—Mangrove area; 6—Normal tectonic faults; 7—Reverse faults; 8—Transforming faults; 9—Volcanic craters and cones; 10—Alluvial fans; 11—Holocene sedimentary plains; 12—Hydrological network. *J.P. Bergoeing, 2013.*

the most ancient volcanic-alluvial fans, predominantly laharics. To the south near the coast are the most modern ones, formed by Holocene landslides that have shaped the Flandrian paleoshoreline (Figure 8.4).

BOCAS DEL TORO ARCHIPELAGO

Bocas del Toro archipelago is made up of 9 islands, 50 keys, and more than 200 islets surrounded by coral reefs. **Colon Island** is the main island of the sector, located northwest of Panama in the Caribbean Sea. With an area of 23.552 mi², it is the largest island in the province of Bocas del Toro and the fourth largest island in Panama. This coral island has built an indoor lagoon, formed between the coast and the insular sector. It is calm, with transparent waters and without waves, enabling the creation of coralline system islands. It is also characterized by a humid tropical climate with two short dry seasons: February-April and September-October. The Caribbean coast of Panama corresponds to a coral formation area, created in parallel with the tectonic raising of the coast. This began in the Miocene and continues today. The oldest formations were fringe and barrier reefs (giving name to the sedimentary geologic formations of Gatun, Veragua, La Yeguada, Sona, Tribique, etc.) attached to the volcanic Talamanca-Tabasara range formation (Tabasara Group formation). The Pliocene-Quaternary orogeny that defined the Panama microplate deformed the sedimentary sector, creating in the lowlands a series of anticline and syncline parallel reliefs following the general northwest-southeast direction of the country, and in altitude the tectonic faulting system was exploited by the topographic network. The Bocas del Toro archipelago coast is of recent formation and can be placed at the beginning of the Quaternary. The Caribbean slope of the Tabasara Mountain Range is extremely steep and subject to frequent landslides, despite the dense tropical jungle that coats it. This sector consists of Secondary and Tertiary geologic formations ranging from Cretaceous to Miocene, where sedimentary formations of limestone, shale, and sandstone are interbedded with lavas, tuffs, and conglomerates. Some granodiorite intrusions of the Miocene complete the geological spectrum of this mountainous area, which is very similar to the Costa Rican Talamanca sector. The accused earrings and heavy alteration of the volcanic-sedimentary material *in situ*, due to the humid tropical environment, explains that despite the important vegetation cover zone, the area is affected by major landslides. Slopes of this sector are characterized by a multiform modeling peculiar to mountainous tropical areas. The Caribbean contact area with the coastal plain of the Tabasara range instead becomes a generalized multi-convex modeling, except in those places where tectonics are very active, forming fault scarps associated with anticlines and synclines.

Meanwhile, the islands comprising Bocas del Toro archipelago are essentially composed of a coral base (Gatun Formation: Miocene limestone, shale, siltstone, and sandstone).On the surface, it is possible to find coated limestone yellow clays decomposed with 70 ft to 100 ft 800 thickness, as well as dolines, some of which are lakes fed by rainwater. The coral is also subject to the karst dissolution phenomenon, creating caves in some sectors, such as **La Virgen cave** on Colon Island. It is possible to say that the basement is affected by a widespread cryptokarst. Living coral are developed both leeward and windward (8% of species), but more species (32%) are developed by the latter. The branched coral *Porites furcata* is the species responsible for 90% of the construction of the more important reefs in shallow waters. Despite the high coral biodiversity and the importance of the livelihood of local residents, most of the reefs are rapidly deteriorating due to overfishing, sediment discharge, and an unplanned increase in tourism. Some sectors, such as north **Bastimentos Island** and the lagoon sector ranging from the **Port of Almirante** to **Laguna Porras**, are composed of outcrops of basaltic-andesitic volcanic rocks and breccias of the Miocene period, which correspond to the Veragua Formation, where corals have also strengthened. All of this material is highly altered in surface, giving equally considerable thickness to clay decomposition. On December 14, 2008, Guillermo Jordan-Garza, a coral expert of the Institute of Marine Sciences and Limnology of Puerto Morales in Mexico (UNAM), published the following in BBC World: *"The situation of coral reefs is severe and is getting worse, by a proliferation of algae due to the warming of sea due to climate change."* He recommended the repopulation of the area with parrot fish that are natural predators of the algae. As we can see, the shoreline archipelago ecosystem is extremely delicate, and any parameter changes may have large negative consequences (Figure 8.5).

The continental sector (the location of the Port of Almirante) is surrounded by mangroves, including the red mangrove (*Rhizophorae mangle*), which is the dominant species in the entire environment and can form monospecific populations in the keys. The other species in the area are salty mangroves or black mangroves (*Avicennia germinans*), white mangroves (*Laguncularia recemosa*), button mangroves (*Conocarpus erecta*), and "piñuela" mangroves (*Pellicera rhizophorae*). They row in scattered form or create small patches,

FIGURE 8.5
Flooded doline or sinkhole by rain effects in Colon Island. *Photo by J.P. Bergoeing, 2009.*

FIGURE 8.6
The Port of Almirante with palafitte houses and incipient mangrove. *Photo by J.P. Bergoeing, 2009.*

especially behind the band of the red mangroves. "Rhizophoraes" mangroves are increasingly reduced by the presence of palafittes (column elevated wood houses), whose residents use the wood for various purposes; however, they are generally in a good state of conservation. The Port of Almirante serves as the main port sector favored by the protected bay and lagoon (Figure 8.6).

In the northern Port of Almirante an extensive marshy coastal plain develops, which is an ancient lagoon exhumed by the positive neotectonics. Farther south is a second lagoon known as **Laguna de Chiriqui**, bounded on the north by the island and peninsula of Boquete, and on the southeast by **Punta Valiente Peninsula**, home of the channel inlet known as **Tigre canal**. This second sector, as well as northern Changuinola, are of recent formation (Middle Quaternary to Holocene) and are formed by landslides covering the old underlying coral formations that are evolving in a cryptokarst.

In general, the area of Flandrian transgression (6500 years B.P.) is present along the coast, reaching the elevation of 10 yd due to the general uplift that continues today. The rivers carry a great deal of water, descending from the Caribbean slope of the Tasabara mountain range. Large meanders form in their lower course because of the rupture of competition, building powerful alluvial fans, as in the case of the **Changuinola River** or the **Guariviara River**. Most flows in deltaic areas are covered by mangroves; the **Cricamola River** is an example of a Mississippi Delta-type construction.

Changuinola River is one of the largest in the area, with a length of 69 mi. Its water network covers a surface of 1236 mi². The most important tributary is the **Teribe River**, and on its shores sit several indigenous communities, such as the Charagare, Doreiyik, and Sieyik. The river is characterized in its lower course by forming a large alluvial fan, probably an old delta today, which was eroded by the same floods of the river. It conditions two of the most depressed sectors, north and south, which were flooded by the Flandrian transgression. The southern sector is the more depressed, holding an old unburied lagoon that extends to **Bocas del Drago** pass. The other noteworthy river is the drainage basin of the Guariviara River (819 mi² and 33 mi in length), where an indigenous community of Ngobe Bugle sits at its shores and remains of anincipient agriculture and fishing, without drinking water or electricity. The Guariviara is a river that suffers catastrophic annual flooding; in its mouth, the river has built a powerful delta cone. It is in this sector that the **Chiriqui Grande port** and **oil terminal** is located, where the inexperience of some operators led to an oil spill in the sea on February 2, 2009 when the "Petrovsk" tanker vessel was racked. It is estimated that about 200,000 gallons spilled, covering an area of 1729 maritime acres, with the inevitable consequences to marine flora and fauna

AZUERO PENINSULA

Azuero Peninsula is the largest of Panama and thus the most southern of Central America. It is characterized by three main well-defined geomorphological areas:

1. **A coastal plain** that does not exceed 32 ft of altitude and extends from the **Gulf of Parita** to **Punta Mala**, following a northwest-southeast direction. It is characterized by a straight coastline of post-Flandrian beaches.
2. **A north hills area** with multi-convex modeling that allows easy contact between the Gulf of Parita and **Montijo Bay**.
3. **A southern mountainous area** averaging 3280 ft of subequal altitudes, sitting between **Alto Garcia** and **Los Yescos**. This last sector is also characterized by a northwest-southeast regional "strike-slip" type faulting system that highlights faults. Azuero Peninsula consists mostly of basalts ranging from Upper Cretaceous to Oligocene, sedimentary rocks of the same age, and intrusive dioritic quartz. This entire complex is an extension of the Ocean Islands' volcanic arc that characterizes Costa Rica from the Nicoya Peninsula to Panama (Weyl, 1980). The mountainous sector also presents mineralization by the effect of hydrothermal alteration (Corral et al., 2009; Figures 8.7 and 8.8).

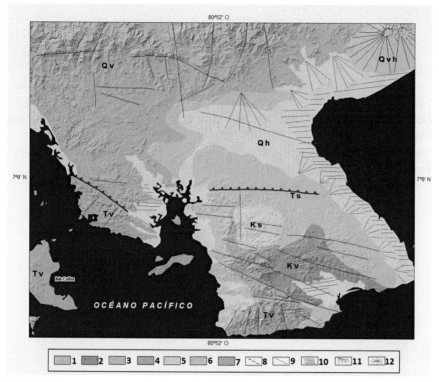

FIGURE 8.7

Azuero Peninsula morphological structural areas: 1—Cretaceous, sedimentary (Ks); 2—Cretaceous, volcanic (Kv); 3—Tertiary, sedimentary (Ts); 4—Tertiary, volcanic (Tv); 5—Quaternary, Holocene, sedimentary (Qh); 6—Quaternary, Holocene, volcanic (QHv); 7—Quaternary, Pleistocene, volcanic (Qv); 8—Reverse faults; 9—Normal faults; 10—Littoral plain; 11—Alluvial fans; 12—Volcanic cones. *J.P. Bergoeing, 2009.*

FIGURE 8.8

Venao Beach, Panama, before being developed as a tourist resort center. Situated on the extreme south of Azuero Peninsula is a perfect natural horseshoe bay. *Photo courtesy of Oceano Community, Cambutal, Panama.*

PANAMA CANAL AREA

The Isthmus of Panama began to consolidate during the Upper Cretaceous period, where basaltic rocks confirming an oceanic volcanic origin prevailed. During the Middle Tertiary, andesitic and intrusive outcrops showed a more continental formation. In the Pliocene, at the end of the Tertiary, the isthmus was shaping up to become the one we know today. It is a sector also crossed by active faulting. It is one of the most depressed parts of Panama, where **Gatun Lake** marks the weaker sector of this part of the country. Folded marine sediments **(Quebrancha syncline)** emerge in the Culebra sector, indicating the tremendous pressure Panama suffers because of the movement of tectonic plates, in particular advancing northward by the pressure exerted by the Nazca plate (Figure 8.9).

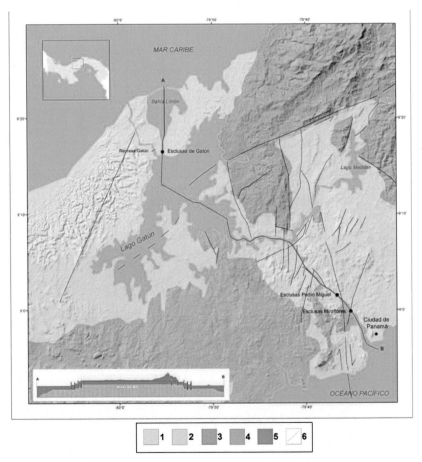

FIGURE 8.9

Structural geomorphology of the Panama Canal sector: 1—Quaternary, Holocene; 2—Quaternary, Pleistocene; 3—Tertiary, volcanic; 4—Tertiary, volcanic and intrusive (basalts and gabbros); 5—Lacustrine and marine areas; 6—Tectonic faults. *J.P. Bergoeing, 2012.*

FIGURE 8.10
Panama Canal floodgate in action. *Photo courtesy of the University of Panama's Department of Geography, 2009.*

If we observe a map of the central Panama sector, we can see that the channel was constructed in a very depressive tectonic zone where flooded lowlands prevailed, characterized by Gatun and Madden Lakes resting in an eminently volcanic basement. Tectonic hazards are particularly present in the Panama Canal because of active faults, including those of **Pedro Miguel** (the earthquake of 1621 caused a displacement of 11 yd between faces of this fault), which is the most important one, as well as Limon and Azota Caballo faults. From a geomorphological point of view, the Panama Canal sector consists of low hill reliefs in the more outstanding parts that have adopted a multi-convex modeling proper to tropical areas, alternating with small volcanic domes standing out in the sector of Panama City and the canal. Topographic forms of different geological formations defined the course of the canal construction started in 1881 by the French company of Ferdinand de Lesseps and ended by Americans in 1914. Panama Canal sovereignty was returned to the Republic of Panama in 1997 with the signing of the Torrijos-Carter Treaty (Figure 8.10).

DARIEN'S GEOMORPHOLOGICAL AREA

The eastern sector of the Panama Canal area includes the **Central depression**, which ranges throughout the eastern area from Bayano Lake (an artificial lake that covers 135 mi^2, possesses a hydroelectric dam, and produces tilapia) to the sector of Darien's border with Colombia. The entire area covers 4.5933 mi^2 and is characterized by three main sectors (Figure 8.11):

1. **An alluvial wavy plain** with the longest rivers in Panama: **Tuira River** (144 mi) and **Chucunaque River** (142 mi extension). It is a Tertiary plain formed by sediments consisting of siltstones, shale, calcareous

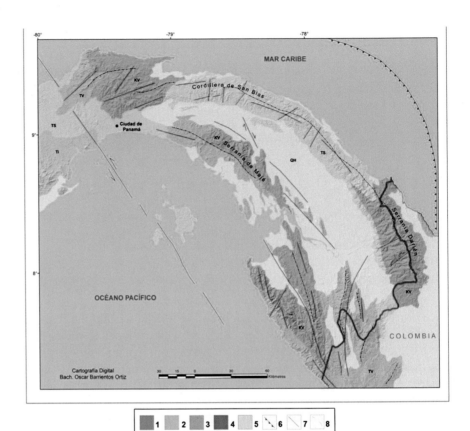

FIGURE 8.11
Structural geomorphology of Darien, Panama: 1—Cretaceous volcanic (Kv); 2—Tertiary sedimentary (Tv);
3—Tertiary volcanic (Tv); 4—Tertiary intrusive (Ti); 5—Quaternary-Holocene; 6—Reverse faults;
7—Normal faults; 8—Watershed line. *J.P. Bergoeing, 2012.*

sandstones, and conglomerates with interbed limestone in thick banks.
It is a much-altered, multi-convex relief. As we approach the bordering
mountains, the relief is accentuated. On the contrary, near the open
coasts where rivers drain, the soil is recent (Holocene), marshy, and is
therefore derived from fluvial and marine sediments.

2. **Two parallel mountain ranges** framing the central depression, formed
   by the **San Blas**, North Darien Mountain Range that dominates the
   Caribbean Sea, whose maximum altitudes are given by **Puna** (4000 ft)
   and **Tacarcuma** hills (5200 ft). To the south rises the **Maje** and **Cañazas**
   ranges, interrupted by the **Tuira River** estuary, and again raising **El
   Sapo**, **Pirre**, **Setetule**, and **Bagre** mountain ranges, extending south in
   the Colombian sector (Figure 8.12).

3. **San Blas archipelago**. Also known as Mulatas archipelago, the San
   Blas extends along Panama's Caribbean coastline from the Gulf of

FIGURE 8.12
Darien's Embera native on the Chagres River. *Photo courtesy of Francesco Taroni, 2013, PanamaViaggi.com.*

FIGURE 8.13
One of the 365 San Blas coral islands, home of Kuna's natives, forming a coral barrier along the southern Caribbean Sea of Panama. *Photo courtesy of Francesco Taroni of PanamaViaggi.com.*

San Blas to the Colombian border. It consists of 365 coral islands and islets. It is home to the Kuna Indians, but only 49 islands are inhabited. On September 7, 1882, the San Blas archipelago was the victim of a tsunami that caused 250 deaths, the product of four giant waves that swept away the surface of the islands. The tsunami occurred as a result of a 7.7-magnitudeunderwater earthquake in the Caribbean facing the islands (Fernandez, 2001; Figure 8.13).

Conclusion

Central America is a region that, in many aspects, has marked contrasts. First of all, its geological formation is very distinctive from north to south. From the Isthmus of Tehuantepec to Northern Nicaragua, we find the oldest formations, ranging from the Paleozoic to the Quaternary. This sector has suffered several orogeneses, concordant with those prevailing in other parts of the planet. From southern Nicaragua to Panama, the isthmus stretches considerably, and it is much younger from a geological point of view going from the Upper Cretaceous to the Quaternary. This is due to the disappearance of the former Farallon plate, by subduction, and its replacement during the Tertiary with the current Caribbean tectonic plate. It is a young, stretched land that emerged particularly during the Miocene and is predominantly volcanic. In fact, the suture between the North American plate (part of the ancient Laurasia continent) and South America (part of the ancient Gondwana continent) only concretized at the end of the Pliocene with the collision of the Caribbean and Cocos tectonic plates.

Like the rest of the planet, Central America has suffered large climate changes characterized by glaciations during the Quaternary period (2 million years). Because of its latitudinal position near the Equator, the glacial and interglacial periods were translated by rhexistasic (dry and erosive) and biostasic (humid and calm) periods, except at the end of the Pleistocene when two glacier parking periods occurred, corresponding to the Illinoian-Riss (200.000–140.000 years) and Wisconsinian-Wurm (90.000–12.000 years) glaciations. These were due to the orogeny in Guatemala (Alto de Cuchumatanes) as in Costa Rica and Panama (Talamanca Mountain Range), where the high peaks rose above 2 mi 300 yd in altitude and left traces as U-valley shapes, small moraines, and tarns or glacial lakes.

Similarly, Central America differs in its reliefs by high plateaus over 1100 yds (Guatemala, Honduras, and Costa Rica) filled by recent volcanic sediments; tectonic and volcanic depressions (El Salvador, Nicaragua, and northern Costa Rica); rivers that have eroded deep river canyons in volcano-sedimentary low consolidated strata; and volcanic collapsed calderas (Retana in Guatemala, Ilopango in El Salvador, Managua in Nicaragua, and Palmares in Costa Rica). Large coastal plains are crossed by rivers that draw large lazy meanders. It is on these coastlines that mangroves have prospered, particularly in the Gulf of Fonseca and the Terraba River mouth in the Pacific Coast. In the Caribbean, large mangroves are present on Honduran and Nicaraguan coastal plains. Currently, mangroves are beginning to suffer anthropic threats from the presence of agro-industries on their borders (banana, pineapple, and oil palm plantations), which are using large amounts of pesticides that are slowly killing mangrove trees. The same applies in some areas to coral reefs that were maintained without significant variations until the twentieth century, but due to the presence of human activities associated with current climate changes, these marine associations have begun to suffer significant reduction.

Central America is also a region that historically was divided into seven independent countries, which today seek integration, with diverse populations and government systems. The establishment of full democracy and economic, industrial, and tourist development are the unique paths of success that should be undertaken to preserve this lovely region.

Bibliography

A.I.D. (USA), 1979. Costa Rica Regional Analysis of Physical Resources. U.S. Army Corps Engineers, AID Resources Inventory Center, Washington, DC, USA.

Agencia Asturiana de Cooperacion, 2009. Ampliación del programa para la regeneración medio ambiental del lago Yojoa, Segunda Etapa. Asturias, España.

Aguilar, T., 1999. Organismos de un arrecife fósil (Oligoceno Superior-Mioceno Inferior) del Caribe de Costa Rica. Rev. Biol. Trop. 47, 453–474.

Alfaro, A., 1913. Rocas volcánicas de Costa Rica. Boletín de Fomento, 123–131.

Alfaro, A., Michaud, G., Biolley, P., 1911. Informe sobre el terremoto de Toro Amarillo, Grecia. Anal. Centro Est. Sismol. de C.R., pp. 35–41. San José, Costa Rica.

Allegre, C.J., Condomines, M., 1976. Fine chronology of volcanic proceses using ^{238}U-^{230}Th systematics. Earth Planet. Sci. Lett. 28, 395–406.

Alvarado, G.E., 2000. Los volcanes de Costa Rica. Editorial Universidad Estatal a Distancia, Costa Rica, 1989.

Alvarado, G.E., Paniagua, S., Tejera, R., 1980. Mapa del vulcanismo Plio-pleistocénico de Costa Rica, 1:700 000. Instituto Geográfico Nacional, Costa Rica.

Alvarado, G.E., Kussmaul, S., Chiesa, S., Gillot, P.-Y., Appel, H., Wörner, G., Rundle, C., 1992. Resumen cronoestratigráfico de las rocas ígneas de Costa Rica, basado en dataciones radiométrica. J. South Am. Earth Sci. 6 (3), 151–158.

Alvarado, G.E., Soto, J.G., Taylor, W.D., 1998. Los flujos piroclàsticos del volcán Arenal del 5 de mayo de 1998 y sus implicaciones para la amenaza de las obras de insfraestructuras cercanas. Bol, OSIVAM 10 (19–20), 1–13.

Alvarado, G.E., Carr, M.J., Turrin, B.D., Swisher, C., Schmincke, H.-U., Hudnut, K.W., 2006. Recent volcanic history of Irazú volcano, Costa Rica: alternation and mixing of two magma batches, implying at least two intracrustal chambers. In: Rose, W.I. Jr. (Eds.), Natural Hazards in Central America. Geol Soc. America Special Paper.

Alvarado, G.E., Pérez, W., Vogel, Th.A., Grüger, H., Patiño, L., 2010. The Cerro Chopo basaltic cone (Costa Rica): an unusual completely reversed graded, pyroclastic cone with abundant low vesiculated cannonball juvenile fragments. J. Volcanol. Geotherm. Res.

Anderson, T.H., 1969. First evidence for glaciation in Sierra Los Cuchumatanes Range, northwestern Guatemala. Geol. Soc. Am. Spec. Pap.121, 387.

Anderson, D.M., 1987. Mapa Geológico de Honduras: Hoja de Lepaterique. Instituto Geográfico Nacional, Tegucigalpa, Honduras, escala 1:50,000.

Arredondo, S., Soto, J.G., 2006. Edad de las lavas del miembro Los bambinos y sumario científico de la Formación Barva, Costa Rica. Revista geológica de América Central 34–35, 59–71.

137

Astorga, A., 1987. El Cretácico Superior y el Paleógeno de la vertiente Pacifica de Nicaragua meridional y de Costa Rica septentrional: origen, evolución y dinámica de las cuencas profundas relacionadas al margen convergente de Centroamérica. Universidad de Costa Rica, 247 pp. Tésis Lic.

Atwater, T., 1970. Implicaciones de las placas tectónicas de la evolución tectónica Cenozoica del oeste de América del Norte. Geol. Soc. Am. Bull. 81, 3513–3556.

Aubrun, Ch.V., 1974. L'Amérique Centrale. Presses Universitaires de France. Que Sais-je N° 513. Paris.

Azambre, B., Tournon, J., 1977. Les intrusions basiques alcalines du Rio Reventazon (Costa Rica). C.R. somm. Soc. Geolo. France, fasc.2, pp. 104–107. Paris. Francia.

Azema, J., Tournon, J., 1980. La péninsule de Santa Elena, Costa Rica. Un massif ultrabasique charié en marge pacifique de l'Amérique Centrale. Note transmise a la séance du 19.11.1979 a l'Academie des Sciences de Paris.

Azema, J., Sornay, J., Tournon, J., 1979. Découverte d'Albien Superieur a ammonites dans le matériel volcano-sédimentaire du Complexe de Nicoya (Province de Guanacaste, Costa Rica. Somm Soc. GeoL de FR. Tome XXI fasc. 3, pp. 129–131.

Azema, J., Glaçon, G., Tournon, J., 1981. Nuevos aportes sobre el Paleoceno con foraminíferos planctónicos del margen Pacífico de Costa Rica. InformeSemestral, pp. 57–69, Jul-Dic, 81. Instituto Geográfico Nacional, Costa Rica.

Barr, K.W., Escalante, G., 1969. Contribución al esclarecimiento de la edad del Complejo de Nicoya, Costa Rica. Pub. Geol. ICAITI 2, pp. 43–47. Costa Rica.

Battistini, R., Bergoeing, J.P., 1980. Observations sur le Quaternaire littoral de la côte Caraïbe du Costa Rica. Revista Quaternaria N° XXII, pp. 237–242, Roma, Italia.

Battistini, R., Bergoeing, J.P., 1982a. Volcanisme recent et variations climatiques Quaternaires au Costa Rica. Bull. Assoc. de Géographes Français, No. 485–486, pp. 96–98, Paris, France.

Battistini, R., Bergoeing, J.P., 1982b. Un exemple de côte a structure faillée quadrillée et Néotectonique active: la côte Pacifique du Costa Rica. Bull. Assoc. de Géographes Français, No. 488, pp. 199–205, Paris, France.

Battistini, R., Bergoeing, J.P., 1983a. Características geomorfológicas del litoral comprendido entre bahía Tamarindo y bahía Culebra, Peninsula de Nicoya, Costa Rica. Instituto Panamericano de Geografía e Historia, Revista Geográfica N° 98, pp. 79–90, México DF, México.

Battistini, R., Bergoeing, J.P., 1983b. Reconnaissance géomorphologique de la façade Pacifique du Costa Rica. Cahiers du CEGET N° 49 3e trimestre 83, pp. 1–73, Bordeaux, France.

Battistini, R., Bergoeing, J.P., 1984. Geomorfología de la costa Caribe de Costa Rica. Instituto panamericano de Geografía e Historia, Revista Geográfica N° 99, pp. 167–188, México DF, México.

Baudez, C.F., et al., 1983. Introducción a la arqueología de Copán, Tomo I Proyecto arqueológico Copán, secretaría de estado en el despacho de Cultura y Turismo, Tegucigalpa D.C., pp. 78–92.

Baxter, S., 1976. Estudio geológico de las Formaciones Mata de Limón y punta Carballo, Costa Rica. Tesis, 55 pp. Escuela Centroamericana de Geología, UCR, Costa Rica.

Beaudet, G., Gabert, P., Bergoeing, J.P., 1982a. Les modelés hérités du Costa Rica. Bull. Assoc. de Géographes Français, No. 488, pp. 183–197, Paris, France.

Beaudet, G., Gabert, P., Bergoeing, J.P., 1982b. La Cordillère de Talamanca et son Piémont (Néotectonique et variations morpho-climatiques dans le Sud-Ouest du Costa Rica. Colloque sur les Piémonts, pp. 121–134. Toulouse 12–15 mai 1982, France.

Bellon, H., Tournon, J., 1978. Contribution de la géochronométrie K/Ar à l'étude du magmatisme du Costa Rica, Amérique Centrale. Bull. Soc. Geol. France 7 (XX), 955–959.

Bergoeing, J.P., 1977. Modelado glaciar en la Cordillera de Talamanca. Informe Semestral, Jul-Dic 77, Instituto Geográfico Nacional, Costa Rica.

Bergoeing, J.P., 1978a. Geomorfología del sector del Cantón de Talamanca, provincia de Limón, Costa Rica 1:200 000. Instituto Geográfico Nacional, Costa Rica.

Bergoeing, J.P., 1978. Geomorfología de Puerto Jiménez, Península de Osa, Costa Rica. Informe Semestral enero-junio 1978, Instituto Geográfico Nacional, San José, Costa Rica.

Bergoeing, J.P., 1978b. Geomorfología de los cursos medio e inferior del río Tempisque, provincia de Guanacaste, Costa Rica 1:200 000. Instituto Geográfico Nacional, Costa Rica.

Bergoeing, J.P., 1979. El volcán Las Nubes. In: Informe Semestral enero-junio 1979, Instituto Geográfico Nacional, San José, Costa Rica.

Bergoeing, J.P., 1982c. Geomorfología de algunos sectores de Costa Rica basada en la fotointerpretación de imágenes del satélite Landsat en la banda espectral MSS 7 (4 mapas a color escala 1:500.000) Informe Semestral, Suplemento Jul-Dic 82, vol. 28, pp. 3–15. Instituto Geográfico Nacional de Costa Rica.

Bergoeing, J.P., 1986. Reconocimiento geomorfológico de la vertiente del Pacífico de Nicaragua. Instituto Panamericano de Geografía e Historia Revista Geográfica 106, 30.

Bergoeing, J.P., 1987a. Photo-interprétation géomorphologique du versant Pacifique duu Nicaragua, Amérique Centrale. Revue Mappe Monde n° 2-1987, pp. 5–8. Montpellier, France.

Bergoeing, J.P., 1987b. Le Costa Rica: contribution à une étude géomorphologique régionale. Tésis de Estado. Universidad de Aix-Marseille II, Francia. 437 pp. (Mirofilmado por la Universidad de Lille).

Bergoeing, J.P., 1987c. L'évolution du Quaternaire au Costa Rica. Cahiers Nantais n° 30-31. Hommage au professeur Gras, pp. 167–187, Nantes. France.

Bergoeing, J.P., 1998. Geomorfología de Costa Rica (croquis, estéreogramas, cartas, fotos). Instituto Geográfico Nacional de Costa Rica, 460 pp.

Bergoeing, J.P., 2000. Geomorfología del Valle de Copán, Honduras (sector comprendido entre Santa Rita y Copán Ruinas). Informe Semestral Instituto Geográfico Nacional de Costa Rica, San José, pp. 47–64.

Bergoeing, J.P., 2005. Le cas du Yellowstone une caldeira exceptionnelle. Revista Geografica N° 137. IPGH, Mexico, Mexico.

Bergoeing, J.P., Protti, Q.R., 2006. Geomorfologia Paleo-Lacustre del sur del lago de Nicaragua. Revista geográfica Internacional Enero 139, 10–38.

Bergoeing, J.P., Protti, R., 2006. Geomorfología Paleo-Lacustre del Sur del Lago de Nicaragua. Revista Geográfica 139, IPGH, México DF. México.

Bergoeing, J.P., 2007. Geomorfologia de Costa Rica. Libreria Francesa, San José, Costa Rica.

Bergoeing, J.P., 2008. Interpretación Geomorfológica del volcán Barú-Panamá. Revista Geográfica IPGH N° 143, México.

Bergoeing, J.P., 2009a. La Trangresión Flandense. Revista Geográfica IPGH N° 144, Mexico.

Bergoeing, J.P., 2009b. Una joven estructura volcánica de Costa Rica, Arenal, volcán turístico, volcán letal. Revista Geográfica, N° 145 IPGH, México DF, México.

Bergoeing, J.P., 2009c. Paisajes volcánicos de Costa Rica. Editorial Jadine. San José. Costa Rica.

Bergoeing, J.P., 2011a. Los dos últimos periodos glaciares y la constitución de sackungs en Talamanca, Costa Rica. Revista Geográfica, No. 149 IPGH, México DF, México.

Bergoeing, J.P., 2011b. Riesgo de desaparición de la flecha litoral de Puntarenas, Costa Rica. Revista Geográfica, N° 149 IPGH, México DF, México.

Bergoeing, J.P., 2012a. El antiguo vulcanismo de la Cordillera de Talamanca. Revista Geográfica, N° 151 IPGH, México DF, México.

Bergoeing, J.P., 2012b. Geomorfología de Isla del Coco, Costa Rica. Revista Geográfica, N° 151 IPGH, México DF, México.

Bergoeing, J.P., 2013. Geografía y civilizaciones antiguas. Revista Geográfica, N° 152 IPGH, México DF, México.

Bergoeing, J.P., Artavia, L.G., 2012. Extensión glaciar y nival durante el riss/illinoiense y el wurm/wisconsiniano en las altas cumbres de Talamanca en el sector fronterizo Costa Rica-Panamá. Revista Geográfica, N° 152 IPGH, México DF, México.

Bergoeing, J.P., Brenes, L.G., 1978a. Geomorfología del cantón de Buenos Aires, provincia de Puntarenas, Costa Rica. 1:200 000. Instituto Geográfico Nacional, Costa Rica.

Bergoeing, J.P., Brenes, L.G., 1978b. Mapa Geomorfológico de Costa Rica 1:1 000 000. Instituto Geográfico Nacional, Costa Rica.

Bergoeing, J.P., Brenes, L.G., 2007. Las calderas concéntricas del Platanar, Costa Rica Revista Geográfica IPGH N° 141 Enero-Junio 2007, México.

Bergoeing, J.P., Herrera, M.M., 2012. El asentamiento precolombino en San Ramón y su imbricación geomorfológica. Jean Pierre Bergoeing y Mauricio Murillo Herrera. Revista Geográfica, N° 152, IPGH, México DF, México.

Bergoeing, J.P., Malavassi, E., 1981a. Síntesis Geológica del valle Central, Costa Rica. 1:100 000 (2 hojas). Instituto Geográfico Nacional, Costa Rica.

Bergoeing, J.P., Malavassi, E., 1981b. Carta Geomorfológica del Valle Central. Escala: 1:50.000 (9 hojas mas texto) editada en colores por Instituto Geográfico Nacional, Costa Rica (900 ejemplares).

Bergoeing, J.P., Protti, R., 2006. Geomorfología Paleo-Lacustre del Sur del Lago de Nicaragua Revista Geográfica, N° 139 IPGH, México DF, México.

Bergoeing, J.P., Protti, M., 2009. Tectónica de placas y sismicidad em Costa Rica. Revista Geográfica del IPGH N° 149 julio-Diciembre 2009, México.

Bergoeing, J.P., Malavassi, E., Protti, R., 1978a. Tres posibles edificios volcánicos del sector Cerros del Aguacate. Informe Semestral jul. Dic. 78. Instituto Geográfico Nacional, Costa Rica.

Bergoeing, J.P., Mora, S., Jimenez, R., 1978b. Evidencias De vulcanismo Plio-Cuaternario en la Fila Costeña, Térraba, Costa Rica. Informe Semestral Jul-Dic 78, Instituto Geográfico Nacional, Costa Rica.

Bergoeing, J.P., Brenes, L.G., Malavassi, E., 1982a. Geomorfología de la hoja Barranca, Costa Rica. Escala 1:50.000 (1 hoja) editada en colores por Instituto Geográfico Nacional, Costa Rica (1.000 exemplaires).

Bergoeing, J.P., Brenes, L.G., Malavassi, E., 1982b. Geomorfología del Pacífico Norte de Costa Rica. Escala: 1:100.000 (11 hojas más texto) editada en colores por Instituto Geográfico Nacional, Costa Rica financiado por CONICIT-US-AID (2.000 ejemplares).

Bergoeing, J.P., Arce, R., Brenes, L.G., Protti, Q., 2007. La caldera de Barbilla, Costa Rica, investigación preliminar. Revista Geográfica IPGH N° 142, México.

Bergoeing, J.P., Brenes, L.G., Fernandez, M.R., 2010a. Las calderas volcánicas de la Cordillera de Talamanca, Costa Rica. Revista Geográfica, N° 148 IPGH, México DF, México.

Bergoeing, J.P., Brenes, L.G., Fernandez, A.M., Ureña, F.M., 2010b. Geomorfología de la cordillera Costeña y de los abanicos aluviales en el piedemonte meridional de la Cordillera de Talamanca. Revista Geográfica, N° 148 IPGH, México DF, México.

Bergoeing, J.P., Arce, R., Brenes, L.G., Protti, R., 2010c. Atlas Geomorfológico del Caribe de Costa Rica. Escala 1:100.000. Editorial SIEDIN Universidad de Costa Rica. 33 pags Color. San José Costa Rica.

Bergoeing, J.P., Brenes, L.G., Salas, D., 2010d. Atlas Geomorfologico de Costa Rica. Escala 1:350.000. Editorial Instituto Costarricense de Electricidad, ICE. San José, Costa Rica.

Berrange, J.P., 1979. Geological map of the Tapanti quadrangle, Costa Rica. Ed. project sponsored by the Ministry of Overseas Development en collaboration with Dirección de Geología, Minas y Petróleo. Ministerio de Economía Industria y Comercio. San José, Costa Rica.

Berry, E.W., 1939. Contribution to the paleobotany of middle and South America. John Hopkins Univ. Stud. en Geol, vol. 13, 168 pp.

Bertrand, M., 1989. Les phénomènes volcaniques et les tremblements de terre de l'Amerique Centrale. Bull. Soc. Geol. France 3ème ser. 27, 494–495.

Bohnberger, O.H., Bengochea, A., Dondoli, C., Marroquin, A., 1966. Report on active volcanoes en Central America during 1957–1965. 11a. Reunión geológica de América Central, Guatemala (Bull Erupt. Tokio 9).

Borgia, A., Poore, C., Carr, M.J., Melson, W.G., Alvarado, G.E., 1988. Structural stratigraphic and petrologic aspects of the Arenal-Chato volcanic system, Costa Rica: evolution of a young stratovolcanic complex. Bull. Volcanol. 50, 86–105.

Boudon, G., Rancon, J.P.H., Kieffer, G., Soto, G., Traineau, H., Rossignol, J.C., 1995. Estilo eruptivo del volcán Rincón de La Vieja; Evidencias de los productos de las erupciones de 1966–70 y 1991–92, Roschildia, Costa Rica. http://www.acguanacaste.ac.cr/rothschildia/v2n2/textos/pag10.html.

Boudon, G., Le Friant, A., Komorowski, J.-C., Deplus, C., Semet, M.P., 2003a. Instabilités des volcans de l'arc Antillais: origine et implications sur les risques volcaniques- Rapport Quadriennal 1999–2002, Comité National Français de Géodésie et Géophysique, CNFGG, XXIIIéme Assemblée Générale de l'UGGI, Saporro (Japon), juillet.

Boudon, G., Semet, M.P., Komorowski, J.-C., Villemant, B., Michel, A., 2003b. Was the last magmatic eruption of la Soufriére, Guadeloupe, in 1440 AD, triggered by partial collapse of the volcano? In: EGS-AGU-EUG Joint Assembly, Nice, France, 06–11 April 2003, Volume 5, Geophysical Research Abstracts, pp. 10398.

Boza, M., 1978. Los Parques Nacionales de Costa Rica. Servicio de Parques Nacionales, San José, Costa Rica.

Brenes-Chaves, G., 1978. Algunas consideraciones sobre posibles problemas biogeográficos en la cuenca del río Sucio. Tesis de Licenciatura en Geografía. UCR, San José, Costa Rica.

Brenes-Monge, M.A., 1968. Contribución a la geología del valle Central Occidental (hoja Turrúcares). Tesis de Ingeniero Agrónomo. UCR, San Jose, Costa Rica.

Brenes-Quesada, L.G., 1976. Análisis geomorfológico de procesos de remoción en masa en parte de la cuenca del río Reventazón, Costa Rica. UCR, San José, Costa Rica.

Brombach, T., Marini, L., Hunziker, J.C., 2000. Geochemistry of the thermal springs and fumaroles of Basse-Terre Island, Guadeloupe, Lesser Antilles. Bull. Volcanol. 61, 477–490.

Brooks, H., 1969. A preliminary report to the Organization of tropical Studies Inc. on Lake Izabal. Geology and Hydrology, 19 pp. USA.

Bullard, F., 1956. Volcanic activity en Costa Rica and Nicaragua en 1954. Trans. Am. Geophys. Union 37, 75–82.

Bullard, F., 1957. Active volcanoes of Central America. XX Congr. Geol. Intern. Ses., vol. 1, pp. 351–371. USA.

Burkart, B., 1978. Offset across the Polochic fault of Guatemala and Chiapas, Mexico. Geology 6, 328–338.

Burkart, B., Deaton, B.C., Dengo, C., Moreno, G., 1987. Tectonic wedges and offset Laramide structure along the Polochic fault of Guatemala and Chiapas, Mexico: reaffirmation of large Neogene displacement. Tectonics 6 (4), 411–422.

Butterlin, J., 1956. La constitution géologique et la structure des Antilles. CNRS, 453 pp. Paris.

Butterlin, J., 1977. Géologie structurale de la région des Caraibes. Masson et Cie, 259 pp. Paris.

Calvert, P.P., 1918. Eruptions of the Costa Rican volcano Irazu en 1917–1918. Proc. Acad. Natl. Sci. Phila, 73.

Camacho, E., 1992. Volcanes en Panamá. Universidad Tecnológica de Panamá – Cepredenac, Instituto de Geociencias, Panama.

Camacho, E., 1997. Los terremotos en el istmo de Panamá. Laboratorio de Geofísica e Hidrología, Universidad de Panamá, Panamá.

Camacho, E., 2007. Seismicity of the Subducted Caribbean Plate in Panama AGU Spring Meeting Abstracts 05/2007.

Camacho Astigarrabía, E., 2009. Sismicidad de las tierras altas de Chiriqui. Tecnociencia 11 (1), Panama.

Carballo, M.A., Fischer, R., 1978. La formación San Miguel. Informe Semestral Ene – Jun 78, pp. 48–144. Instituto Geográfico Nacional, Costa Rica.

Carlut, J., Quidelleur, X., 2000. Absolute paleointensities recorded during the Brunhes chron at La Guadeloupe Island. Phys. Earth Planet. Inter. 120, 255–269.

Carlut, J., Quidelleur, X., Courtillot, V., Boudon, G., 2000. Paleomagnetic directions and K/Ar dating of 0 to 1 Ma old lava flows from La Guadeloupe Island (French West Indies): implications for time averaged field models. J. Geophys. Res. 105 (B1), 835–849.

Carr, M.J., 1984. Symmetrical and segmented variations of physical and geochemical characteristics of the Central American volcanic front. J. Volcanol. Geotherm. Res. 20, 231–252.

Carr, M.J., Craig, A., Bruce, G., 1986. Nuevos análisis de lavas y bombas del Rincón de La Vieja, Costa Rica. – Bol. de Vulcanología, vol. 16, pp. 23–30. Heredia, Costa Rica.

Castillo, R., Kruschensky, R., 1977. Geologic map and cross section of the Abra quadrangle, Costa Rica. Misc. Invest. Series U.S. Geological Survey Map 1-992, Arlington, USA.

Chaves, R., 1969. Características físicas, químicas y minerológicas de los materiales eruptados por el volcán Arenal. Informe Semestral Ene-Jun 69. Instituto Geográfico Nacional, Costa Rica.

Chaves, R., Saenz, R., 1974. Informe técnico y geológico de la cordillera de Tilarán. Ministerio de Econ. Ind. y Com Dirección Geol. Minas y Petrol. San José, Costa Rica.

Chiesa, S., Civelli, G., Gillot, P.-Y., Mora, O., Alvarado, G.E., 1992. Rocas piroclásticas asociadas con la formación de la caldera de Guayabo, cordillera de Guanacaste, Costa Rica. Rev. Geol. América Central 14, 59–75.

Coates, A.G., Obando, J.A., 1996. The geologic evolution of Central American Isthmus. In: Jacson, J.B.C., Budd, A.F., Coates, A.G. (Eds.), Evolution an Environment in Tropical America. Univ. Chicago Press, USA.

Corral, I., Cardellach, E., Gómez-Gras, D., Canals, A., 2009. Contribución al conocimiento de la geología del depósito de Au-Cu de "La Pava" (Península de Azuero, Panamá) Contribution to the knowledge of the geology of the "La Pava" Au-Cu deposit (Azuero Peninsula, Panama) GEOGACETA, 46. Sociedad Geológica de España, Barcelona.

Crossland, C., 1927. The expedition of the south Pacific of S.S.St.George Marine Ecology and coral formations in the Panama region, the Galapagos and Marquises islands. Royal Society of Edinburgh.

Cuffey, K., Marshall, S., 2000. Substantial contribution to sea-level rise during the last interglacial from the Greenland ice sheet. Nature 404, 591–594.

de Boer, J., 1974. Mapa geofísico preliminar de Costa Rica. IGN-CR. San José, Costa Rica.

de Boer, J.Z., Defant, M.J., Stewart, R.H., Restrepo, J.F., Clark, L.F., Ramirez, A.H., 1988. Quaternary calc-alkaline volcanism in western Panama: regional variation and implication for the plate tectonic framework. J. South Amer. Earth Sci. 1, 275–293.

DeMets, C., Gordon, R.G., Argus, D.F., Stein, S., 1990. Current plate motions. Geophys. J. Int. 101, 425–478.

Delmelle, P., Stix, J., Baxter, P.J., García-Àlvarez, J., Barquero, J., 2002. Atmospheric dispersion, environmental effects and potential health hazard associated with the low-altitude gas plume of Masaya volcano, Nicaragua. Bull. Volcanol. http://dx.doi.org/10.1007/s00445-002-0221-6.

Dengo, G., 1959. Bibliografía de la geología de Costa Rica. UCR Depto. Public. Serv. Cienc. Natur. vol. 3, 27 pp. Costa Rica.

Dengo, G., 1960. Notas sobre la geología de la parte central del litoral Pacífico de Costa Rica. Informe Semestral Jul-Dic 60, pp. 43–58. Instituto Geográfico Nacional, Costa Rica.

Dengo, G., 1961. Notas sobre la geología de la parte central del litoral Pacifico de Costa Rica II. Informe Semestral Jul-Dic 61, pp. 43–63. Instituto Geográfico Nacional, Costa Rica.

Dengo, G., 1962a. Tectonic-igneous sequence in Costa Rica. Petrol. Studies Vol. en honor of A Buddington Geol. Soc. Amr., pp. 133–161, USA.

Dengo, G., 1962b. Estudio Geológico de la región de Guanacaste, Costa Rica. Instituto geográfico Nacional, 112 pp. San José, Costa Rica.

Dengo, G., 1962c. Mapa geológico generalizado de la provincia de Guanacaste y zonas adyacentes. Preparado con base en el mapa IGN y en los levantamientos geológicos efectuados por la Compañía Petrolera de C.R. por los geólogos Dion; L. Gadner et al. Instituto Geográfico Nacional, Costa Rica.

Dengo, G., 1967. Geological structure of Central America. In: Studies in Tropical Oceanography Proc. Intern. Conf. on Trop. oc. Univ. of Miami, USA, pp. 56–73.

Dengo, G., 1968. Estructura Geológica, Historia Tectónica y Morfología de América Central. Centro regional de ayuda técnica AID, México.

Dengo, G., 1969. Problems of tectonic relations between Central América and the Caribbean. Trans. Gulf Coast Assoc. Geol. Soc. 19, 3121–3320.

Dengo, G., 1973. Estructura Geológica, Historia tectónica y Morfología de América Central. Instituto Centroamericano de Investigaciones y tecnología Industrial ICAITI, Costa Rica.

Dengo, G., Chaverri, O.H., 1951. Reseña geológica de la región S.O. de la meseta Central de Costa Rica. Revista Universidad de Costa Rica 5, 313–326.

Dengo, C., Dengo, G., 1985. Posible unión de Fallas Polochic y Motagua en el occidente de Guatemala. http://desastres.usac.edu.gt/documentos/pdf/spa/doc7545/doc7545-contenido.pdf.

Dengo, G., Bohnberger, O.H., Bonis, S. 1970. Tectonics and volcanism along the Pacific et marginal zone of Central América. Geol. Rdch. Stuttgart, vol. 59, pp. 1215–1232. RFA.

Denyer, P., Kussmaul, S., 2000. Geología de Costa Rica. Editorial Tecnológica de Costa Rica, 515 pp. Cartago, Costa Rica.

Diaz, R.A., 1981. Magnitud de un sismo generado por fallas ubicadas en el valle Central. Tesis de Licenciatura, Facultad de Ingeniería UCR, Costa Rica.

Dondoli, C., 1943a. La región de El General. Condiciones geológicas y geoagronómicas de la zona. Depto. Nac. de Agric. BoL Tec. 44, pp. 1–16. San Pedro Montes de Oca, Costa Rica.

Dondoli, C., 1943b. Visión rápida geoagronómica de la meseta Central. Depto. Nac. de Agric. Bol. Tec. 45, San Pedro M.O. Costa Rica.

Dondoli, C., 1950. Liberia y sus alrededores. Suelo Tico 4, pp. 18–19. San Josê, Costa Rica.

Dondoli, C., 1951a. Observaciones sobre las andesitas*. del Virilla encontradas en la perforación del túnel Ing. Dengo y Chaverri S., pp. 324–325. Costa Rica.

Dondoli, C., 1951b. Zona de Palmares; estudio geoagronómico. Bol. Tec. 5, 16 pp. Ministerio Agric. e Industria. Costa Rica.

Dondoli, C., 1965. Vulcanismo reciente en Costa Rica. Dir. Geol. Minas y Petrol. Minist. Ind. 16 pp, Costa Rica.

Dondoli, C., 1970. Localización de un horizonte laterítico-bauxítico en la zona de Cartago concreciones de gibsita en la laterítica bauxíica de Birrisito de Cartago. Dir. Gral. Geol. Minas y Petro. Informes Tec. y Notas Geol. No. 36, pp. 12, UCR C.R.

Dondoli, C., Chaves, R., 1968. Mapa adjunto al estudio geológico del valle Central. Dir. Geol. Minas y Petrol. Costa Rica.

Dondoli, C., Torres, M.A., 1954. Estudio geológico y mineralógico pedagógico de la región oriental de la meseta Central. Ministerio de Agricultura e Industrias, 180 pp. Costa Rica.

Dondoli, C., Dengo, G., Malavassi, E., 1968. Mapa Geológico de Costa Rica. 1:700 000 Edic. Prel. Dir. Geol. Minas y Petrol. Costa Rica.

Donovan, S.K., Jackson, T.A. (Eds.), 1994. Caribbean Geology: An Introduction. University of the West Indies Publisher's Association, Kingston, Jamaica, pp. 13–39, QE 220 C34.

Echandi, E., 1981. Unidades volcánicas de vertiente norte de la cuenca del río Virilla. Escuela Centroamericana de Geología. Tesis., 123 pp. UCR, Costa Rica.

Elizondo, C., Bergoeing, J.P., 1980. Seasat radar data interpretation for geomorphological research and mapping in Costa Rica (Summary). In: For the Fourteenth Int. Symp. on Remote Sensing of Environment, April 23–30, San José, Costa Rica. Ann Harbor, USA.

Escalante, G., 1965. Mapa geológico de la parte superior de la cuenca del río Reventazón, Costa Rica. 1:500 000 Instituto geográfico Nacional, Costa Rica.

Escalante, G., 1968. Mapa geológico preliminar de la región sureste de Costa Rica. La cordillera de Talamanca y sectores próximos. 1:300 000 inedit.

Espinoza, E., Gutiérrez, C., Cerrato, D., Vázquez-Prada, D., 2008. Cartografía Geológica y Geomorfológica de la Reserva Natural Laguna de Apoyo. Programa Integral por le Ordenamiento Ambiental de Apoyo – AMICTLAN-Geólogos.

Fan, G.W., Beck, S.L., Wallace, T.C., 1993. The seismic source parameters of the 1991 Costa Rica aftershock sequence: evidence for a transcurrent plate boundary. J. Geophys. Res. 98, 15,759–15,778.

Felis, T., et al., 2004. Increased seasonality in Middle East temperatures during the last interglacial period. Nature 249, 164.

Fernandez, M., 2001. Daños, efectos y amenazas de tsunamis en América Central. Universidad de Costa Rica, Centro de investigaciones Geofísicas.

Fernandez, M., Rodriguez, R., 1952. Estudio geológico del cantón de Atenas. Tesis de Licenciatura. Escuela de Geología, UCR, Costa Rica.

Feuillet, N., 2000. Sismotectonique des Petites Antilles. Liaison entre activité sismique et volcanique. Thése de Doctorat, Université de Paris 7 René Diderot, pp. 1–283.

Feuillet, N., Manighetti, I., Tapponnier, P., 2001. Extension active perpendiculaire à la subduction dans l'arc des Petites Antilles (Guadeloupe, Antilles Françaises). C.R. Acad. Sci. Paris, Sciences de la Terre et des Planètes 333, 583–590.

Feuillet, N., Tapponnier, P., Manighetti, I., Villemant, B., King, G.C.P., 2004. Differential uplift and tilt of Pleistocene reef platforms and Quaternary slip rate on the Morne-Piton normal fault (Guadeloupe, French West Indies). J. Geophys. Res. 109, B02404. http://dx.doi.org/10.1029/2003JB002496, 18 pp.

Field, M.J., 2001. Sea levels are rising. Pacific Magazine (December).

Fisher, D.M., Gardner, T.W., Marshall, J.S., Montero, W., 1994. Kinematics associated with late Tertiary and Quaternary deformation in Central Costa Rica: western boundary of the Panamá microplate. Geology 22, 263–266.

Franco, J.C., 1978. La formación Coris, (Mioceno, valle Central de Costa Rica). Tesis de Licenciatura. Escuela Centroamericana de Geología, UCR, Costa Rica.

Franco, A., 2008. Cinématique Actuelle du Nord de l'Amérique Centrale: Zone de Jonction Triple Amérique du Nord Amérique-Cocos-Caraïbe. Apport des données sismologiques et géodésiques aux modèles régionaux. Université Paris Sud – Paris XI.

Frogley, M.R., et al., 1999. Climate variability in Northwest Greece during the last interglacial. Science 285, 1886–1888.

Gillot, P.Y., Chiesa, S., Alvarado, G.E., 1994. Chronostratigraphy and evolution of the Neogene-Quaternary volcanism in north Costa Rica: the Arenal volcano-structural frame work. Revista Geológica de America Central. 17, 45–53.

Girod, M., 1978. Les roches volcaniques, Petrologie et cadre structural. Doin Editeurs, Paris, 239 pp.

Goes, S.D.B., Velasco, A.A., Schwartz, S., Lay, Y.T., 1991. The April 22, 1991, Valle de la Estrella, Costa Rica (Mw=7.7) earthquake and its tectonic implications: a broadband seismic study. J. Geophys. Res. 98, 8127–8142.

Gomez, L.D., Valerio, C.E., 1971. Lista preliminar ilustrada de los moluscos fósiles de la formación río Banano (Mioceno), Limón Costa Rica. Informe Semestral Ene-Jun 71, Instituto Geográfico Nacional, Costa Rica.

Granados, R., 1979. Proyecto geotérmico, investigación geológica en la zona de la Caldera de Guayabo y alrededores. Inst. Cost. Electricidad. Informe Geol., 45 pp. Costa Rica.

Güendel, F., Pacheco, J., 1992. The 1990–1991 seismic sequence across central Costa Rica: evidence for the existence of a micro-plate boundary connecting the Panama deformed belt and the Middle America trench. EOS Trans. Am. Geophys. Un. 73, 399.

Guilcher, A., 1966. Les grandes falaises et mégafalaises des côtes sud-ouest de l'Irlande. Ann. Géograph. 407, 26–38.

Guilcher, A., Berthois, L., Battistini, R., 1962. Formes de corrosion littorale dans les roches volcaniques, particulierement de Madagascar et au Cap vert (Senegal). Cahiers Océanographiques XVIe année 4, Paris.

Gutenberg, B., Richter, C.H.F., 1942. Seismicity of Central and south América. Geol. Sci. 4, 455.

Gutierrez, B.F., 1955. Expedición del doctor Richard Weyl al Macizo del Chirripó, Bosquejo Geológico de la cordillera de Talamanca. Instituto Geográfico Nacional, Costa Rica.

Gutierrez, D., 1962. Apuntes sobre un viaje al volcán Tenorio. Informe Semestral Ene-Jun 962, pp. 27–30. Instituto Geográfico Nacional, Costa Rica.

Gutierrez, B.F., 1963a. Actividad volcánica del Irazú. Informe Semestral Ene-Jun 63, pp. 33–38. Instituto Geográfico Nacional, Costa Rica.

Gutierrez, B.F., 1963b. Hallazgo de resto de un mamut. Informe Semestral Ene-Jun 63, pp. 41–47. Instituto Geográfico Nacional, Costa Rica.

Gutsche, A., 2005. Distribution and habitat utilization of "Ctenosaura bakeri" in Utila. Iguana Review 12 (3).

Haas, O., 1942. Miocene molluscs from Costa Rica. J. Paleontol. 16, 307–316.

Hall, C., 1984. Costa Rica, una interpretación Geográfica con perspectiva histórica. Editorial Costa Rica, 467 pp. San José, Costa Rica.

Harrison, J.V., 1953. The geology of the Santa Elena península en Costa Rica, Central América. In: Proc. Seventh Scient. Cong., vol. 2, USA, pp. 102–114.

Harrison, J.V., 1956. East and West lines en Costa Rica. In: XX Congreso Geológico Interamericano, Resum pag 281–282, México.

Hastenrath, S., 1973. On the Pleistocene glaciation of the Cordillera de Talamanca, Costa Rica. Zeitschrift fur Gletscherkunde und Glazial-geologie. 9, 105–121. DFR

Hastenrath, S., 1974. Spuren Pleistozäner vereigsung in den Altos de Cuchumatanes, Guatemala. Eiszeïltalter U. Gegenwert, vol. 5, pp. 25–34. R.F.A.

Hastenrath, S., 1974. Spuren pleistozaener Vereisung in den Altos de Cuchumatanes, Guatemala. Traces of Pleistocene glaciation in the Altos de Cuchumatanes. Eiszeit. Gegenw. 25, 25–34 DFR.

Hazlett, R.W., 1977. Geology and hazard of the San Cristóbal volcanic complex – Nicaragua. Dartmouth College, Hanover New Hampshire, (Tesis de Maestría) – June.

Healy, J., 1969. Notas sobre los volcanes de la sierra Volcánica de Guanacaste, Costa Rica. Informe Semestral Ene-Jun 69, pp. 37–47. Instituto Geográfico Nacional, Costa Rica.

Henningsen, D., 1964. Estratigrafía y paleogeografía de los sedimentos del Cretáceo Superior y del Terciario, en el sector Suresre de Costa Rica. Informe Semestral Jul-Dic 64. Instituto Geográfico Nacional, Costa Rica.

Hernitz, S.R., 1981. Landforms under a tropical wet forest cover on the Península Osa, Costa Rica. Z. Geomorphol. 25, 259–270.

Hey, R., 1977. Tectonic evolution of the Cocos-Nazca spreading. Geol. Soc. Am. Bull. V. 88, 1404–1420.

Hey, R., Johnson, G.L., Lawrie, A., 1977. Recent plate motions in the Galapagos area. Geol. Soc. Bull. V. 88, 1385–1403.

Hoffstetter, R., et al., 1960. Lexique stratigraphique international. Vol. 5 Amérique Latine: Fasc. 2 Amérique Centrale, 368 pp, 8 cartes, CNRS, Paris.

Holdridge, L., 1969. Ecología basada en zonas de vida. Instituto Interamericano de Ciencias agrícolas IICA, Turrialba, Costa Rica.

Horn, S.P., 1990. Timing of deglaciation in the Cordillera de Talamanca, Costa Rica. Climate Res. 1, 81–83.

Horn, S.P., 1993. Postglacial vegetation and fire history in the Chirripó Páramo of Costa Rica. Quatern. Res. 40, 107–116.

Incer-Barquero, J., 1980. Los Cráteres del volcán Masaya. In: Maria, M.R. (Ed.), Boletín Nicaragüense de Bibliografía y Documentación, vol. 35. pp. 1–34. Managua, Nicaragua.

Instituto Costarricense de Electricidad (ICE), 1976a. Mapa geológico regional (Guanacaste) 1:400 000. Departamento de Geología, ICE, Costa Rica.

Instituto Costarricense de Electricidad (ICE), 1976b. Mapa Geológico regional (proyecto geotérmico) 1:400 000. Depto. Geol. ICE, San Josê, Costa Rica.

Instituto Costarricense de Electricidad (ICE), 1976c. Boletín Hidrológico 10, pp. 349. San José, Costa Rica.

Instituto Interamericano de Ciencias Agricolas, 1969. Inventario de recursos de los cantones Atenas, Esparta, Orotina y San Mateo, Costa Rica. 14 pp, IICA, Turrialba, Costa Rica.

Instituto Meteorologico Nacional, 1986. Data climática de Costa Rica 1888–1986. Ministerio de Agricultura y Comercio, San José, Costa Rica.

Instituto Nicaraguense de Estudios Territoriales. Ineter, 2002. Actualización del Mapa de fallas geológicas de Managua. Informe Técnico, Managua, Nicaragua, Abril 2002.

Instituto Panamericano de Geografía e Historia IPGH, n.d. Atlas climatológico e hidrológico del istmo centroamericano. Publicación 367 IPGH, Guatemala.

Jacob, K.H., Pacheco, Y.J., 1991. The M-7.4 Costa Rica earthquake of April 22, 1991, tectonic setting, teleseismic data, and consequences for seismic hazard assessment. Earthquake Spectra 7B.

Jager, G., 1977. Geología de las mineralizaciones de cromita al este de la península de Santa Elena, provincia de Guanacaste, Costa Rica. Tesis de grado, Escuela Centroamericana de Geología UCR Costa Rica, 136 pp.

James, K.H., 1998. A Simple Synthesis of Caribbean Geology. PDF of Keith James' Caribbean plate model.

Kaspar, F., et al., 2005. A model-data comparison of European temperatures in the Eemian interglacial. Geophys. Res. Lett. 32, L11703.

Komorowski, J.-C., 2003. Diversité du volcanisme Terrestre: processus et produits. In: De Wever, P. (Ed.), Le volcanisme: cause de mort et source de vie. Editions Vuibert, Paris, pp. 27–112.

Komorowski, J.-C., Boudon, G., Semet, M., Coudret, E., Villemant, B., Le Friant, A., 2004. A new look at the pyroclastic eruptive history of Soufrière of Guadeloupe (French West Indies) in the past 50,000 years: implications for GIS-based hazard scenario definition. IAVCEI General Assembly 2004, Pucon, Chili, November 14–19, abstract.

Kruschensky, R., 1972. Geology of the Istaru Quadrangle, Costa Rica. Geol. Surv. Bull. 1358, 46.

Kruschensky, R., Malavassi, E., Castillo, R., 1976. Mapa de reconocimiento geológico y cortes transversales de Costa Rica. Oficina de Defensa Civil y Direc. Geol. minas y Petrol Minist. de Ind. y Com Costa Rica.

Kueny, J.A., Day, M.J., 2002. Designation of protected karstlands in Central America: a regional assessment. J. Cave Karst Stud. 64 (3), 165–174.

Kukla, G., 2000. The last interglacial. Science 287, 987–988.

Kussmaul, S., Paniagua, S., Gainza, J., 1982. Recopilación, clasificación e interpretación petroquímica de las rocas de Costa Rica. Informe Semestral Jul-Dic, vol. 82, pp. 17–79. Instituto Geográfico Nacional, Costa Rica.

Kuyjpers, E.P., 1980. La Geología del Complejo Ofiolítico de Nicoya, Costa Rica. Informe Semestral Ene-Jun, vol. 80. Instituto Geográfico Nacional, Costa Rica.

Lachniet, S.M., Roy, J.A., 2011. Quaternary glaciations—extent and chronology a closer look. In: Developments in Quaternary Science. Elsevier, Costa Rica and Guatemala.

Lachniet, M.S., Seltzer, G.O., 2002. Late Quaternary glaciation of Costa Rica. Geol. Soc. Am. Bull. 114 (5), 547–558. USA.

Laserre, G., 1975. América media. México, América Central, Antillas Guayanas. Editorial Ariel, Barcelona.

Lea, D., et al., 2000. Climate impact of late Quaternary equatorial Pacific sea surface temperature variations. Science 289, 1719–1723.

Lépine, J.-C., Hirn, A., Komorowski, J.-C., 2001. Sismicité de la Soufriére de Guadeloupe, Atelier sur les aléas volcaniques – Les volcans antillais: des processus aux signaux. In: Programme National des Risques Naturels – PNRN (CNRS) – Institut National des Sciences de l'Univers – INSU, BRGM, CEA, CEMAGREF, CNES, IRD. 18–19 janvier 2001, Paris, abstract volume, p. 24.

Leyrit, H., 2003. Caractéristiques des volcans et provinces volcaniques de France. In: De Wever, P. (Ed.), Le volcanisme: cause de mort et source de vie. Editions Vuibert, Paris, pp. 143–202.

Linares, O.F., 1975. Prehistoric agriculture in tropical highlands. Science 187.

Linkimer, L., 2003. Neotectónica del extremo oriental del Cinturón Deformado del Centro de Costa Rica. Tesis de Licenciatura, Universidad de Costa Rica, Escuela de Geología.

Lloyd, J., 1963. Historia Tectónica del orogeno Sur Centroamericano. Informe Semestral Ene-Jun, vol. 63, pp. 67–96. Instituto Hgeográfico Nacional, Costa Rica.

Lockwood, J.P., Benumof, B.T., 2000. Geologic investigations field report, Isla Del Coco, Costa Rica. Prepared for the Government of Costa Rica and the Puffin Investment Company, Ltd., p. 29.

López de Velasco 1570, 1993. (Revista Conservadora del Pensamiento Centroaméricano, Vol. 24 N° 121, Octubre 1970.) In: Feldman, L.H. (Ed.), Mountains of Fire, Land That Shake. Labeyrinthos, California, USA.

Lopez Ramos, E., 1975. Geological summary of the Yucatan Peninsula. In: Nairn, A.E.M., et al. (Eds.), The Gulf of Mexico and the Caribbean. Plenum Press, New York, pp. 257–282.

Lorenz, V., 1973. On the formation of maars. Bull. Volcanol. 37, 183–204.

Lovell, W.G., 2005. Conquest and Survival in Colonial Guatemala, third ed. McGill-Queens University Press, Canada.

MacMillan, A.I., Gansa, P.B., Alvarado, I.G., 2004. Middle Miocene to present plate tectonic history of the southern Central American Volcanic Arc. Tectonophysics 392, 325–348.

Madrigal, R., 1972. Resumen de la estratigrafía de Costa Rica. Escuela Centroamericana de Geología, UCR, Costa Rica.

Madrigal, R., 1977. Evidencias geomorfológicas de movimientos tectónicos recientes en el valle de El General. Revista Ciencia y Tecnología. UCR, Costa Rica, pp. 97–106.

Madrigal, R., 1978. Terrazas marinas y tectonismo en península de Osa, Costa Rica. Instituto Panamericano de Geografía e Historia, Revista Geográfica 85, México DF, México.

Madrigal, R., Malavassi, E., 1967a. Reseña geológica del Area Metropolitana. Inf. Tecn. y Notas Geol., No. 29, 9 pp. Dir. Gener. de GeoL Minas y PetroL Costa Rica.

Madrigal, R., Malavassi, E., 1967b. Reseña geológica del área metropolitana. Inf. Tecn y Notas Geol., No. 29, 9 pp. Dir. Gen. Geol. Minas y Petrol. Costa Rica.

Malavassi, E., 1960. Algunas localidades de Costa Rica con foraminíferos grandes. Informes. ciudad Universitaria Anno 1, Costa Rica.

Malavassi, E., 1962. Nota geológica preliminar sobre la región de Tilarán, entre la laguna Cote y Cuevas de Venado. Inf. Depto. de GeoL Minas y Petrol., Costa Rica, 9 pp.

Malavassi, E., 1965. Reseña Geológica del Valle Central de Costa Rica. Inf. Técn. Y Notas Geol. No. 4, pp. 14–22. UCR, Costa Rica.

Malavassi, E., 1967. Reseña Geológica de la zona de Turrialba. UCR Dirc. Geol. Minas y Petrol. Ano 6, pp. 9–35, Costa Rica.

Malavassi, E., 1975. Explicación de un perfil geológico a través de Costa Rica. Revista Geográfica de América Central 3, 27–31.

Malavassi, E., Chaves, R., 1970. Estudio geológico regional de la zona Atlántico norte de Costa Rica. Inf. Tecn. y Notas Geol. Ano 9, 35 pp, UCR, Costa Rica.

Malavassi, E., Madrigal, R., 1967. Mapa geológico del área metropolitana de Costa Rica. Dir. de Geol. Minas y Petrol. Min. de Ind. y com Costa Rica.

Malavassi, E., Madrigal, R., 1970. Reconocimiento geológico de la zona norte de Costa Rica. Dir. de Geol. Minas y Petrol Ano 9, 38 pp, Costa Rica.

Malfait, B., Dinkelman, M., 1972. Circum-Caribbean tectonic and igneous activity and the evolution of the Caribbean plate. Geol. Soc. Am. Bull. 83 (2), 251–271.

Mann, P., 1999. Caribbean sedimentary basins: classification and tectonic setting from Jurassic to present. In: Mann, P. (Ed.), Caribbean Basins. In: Hsu, K. (Ed.), Sedimentary Basins of the World Series, vol. 4. Elsevier Science B.V., Amsterdam, The Netherlands, pp. 3–31.

Mapa Geológico de la República de Panamá, 1991. Escala 1:250.000. Ministerio de Comercio e Industria. Instituto Geográfico Nacional Tommy Guardia. Panamá.

Marko, P.B., Jakson, J.B.C., 2001. Patterns of morphological diversity among and within arcid bivalve species pairs separated by the Isthmus of Panama. J. Paleontol. 590–606.

Marshall, J.S., Fischer, D.M., Gadner, T.W., 1993. Western margin of the Panama microplate, Costa Rica: kinematics of faulting along a diffuse plate boundary. Geol. Soc. Am. Abstr. Progr. 25, A-284.

Martrat, B., et al., 2004. Abrupt temperature changes in the Western Mediterranean over the past 250,000 years. Science 306, 1762–1765.

McCulloch, M., et al., 1999. Coral record of equatorial sea-surface temperatures during the penultimate deglaciation at Huon Peninsula. Science 283, 202.

McNeill, D.F., Coates, A.G., Budd, A.F., Borne, P.F., 2000. Integrated paleontologic and paleomagnetic stratigraphy of the upper Neogene deposits around Limon, Costa Rica: a coastal emergence record of the Central American Isthmus. Geol. Soc. Am. Bull., 963–981.

Medina, M.C.E., 2009. Modelos numéricos y teledetección en el lago de Izabal, Guatemala. Tesis de doctorado Universidad de Cádiz, España.

Mesa, T., 1979. Consideraciones generales sobre la morfoestructura y el modelado climático de los Cerros de La Carpintera y su relación con el conjunto Irazú, Costa Rica. Tesis de Licenciatura en Geografía, UCR, Costa Rica.

Montero, W., 1975. Estratigrafía del Cenozoico del área de Turrucares, provincia de Alajuela, Costa Rica. Tesis de Bachillerato, Escuela Centroamericana de Geología UCR Costa Rica, 40 pp.

Mora, S., 1977. Estudio geológico del cerro Chopo. Revista Geográfica de América Central 5–6.

Mora, S., 1978a. Estudio geológico de los Cerros de Barra Honda y alrededores. Tesis de Bachillerato, 173 pp. Escuela Centroamericana de Geología. UCR, Costa Rica.

Mora, S., 1978b. Proyecto Talamanca. Informe de reconocimiento geológico de los sitios de presa Bri-Bri y Sheuab, Depto de Geol. ICE, Costa Rica, 35 pp.

Mora, S., 1979a. Mapa Geológico de la región Sureste del valle de El General 1:500 000 Escuela Centroamericana de Geología.

Mora, S., 1979b. Estudio geológico de la región sureste del valle de El General. Tesis de Licenciatura, 2 Tomos. Escuela Centroamericana de Geología, UCR, Costa Rica.

Mora, S., 1981. Clasificación Morfotectónica de Costa Rica. Informe Semestral Jul-Dic, vol. 81, 26 pp. Instituto Geográfico Nacional, Costa Rica.

Mullerried, F., 1957. La geología de Chiapas, Gobierno Constitucional del estado de Chiapas, 180 pp.

Murata, K.J., Dondoli, C., Saenz, R., 1966. The 1963–1965 Eruption of Irazú volcano, Costa Rica. Bulletin 29, 765–796.

Newman, A.V., Schwartz, S.Y., Gonzalez, V., De Shon, H.R., Protti, J.M., Dorman, L.M., 2002. Along-strike variability in the seismogenic zone below Nicoya Peninsula, Costa Rica. Geograph. Res. Lett 29 (20–77). http://dx.doi.org/10.1029/2002 GL 015409.

Nuhn, H., 1973. Regionalización de Costa Rica para la planificación del desarrollo y la administración. OFIPLAN, Costa Rica.

Olsson, A., 1922. The Miocene of northern Costa Rica. Part 1–2. Bull. Amer. Paleont. 9 (39), 179–460.

Oviedo-Padrón, E.G., 2005. Análisis geológico-estructural del complejo de maares de Valle de Santiago, Campo Volcánico Michoacán-Guanajuato, México. Tesis profesional, Universidad Autónoma de Nuevo León, Facultad de Ciencias de la Tierra, Linares, N.L., p. 119.

Pacheco, J.F., Quintero, R., Vega, F., Segura, J., Jimenez, W., González, V., 2006. The Damas (Mw 6.4) Costa Rica, earthquake of November 20, 2004: aftershocks and slip distribution. Bull. Seim. Soc. Am. 96, 1332–1343.

Paniagua, S., 1984. Contribución al conocimiento de la geología y petrología del vulcanismo Plio-pleistocénico de la cordillera Central de Costa Rica. Tesis de Maestría, Universidad de Chile, Chile.

Pérez, W., 2000. Vulcanología y petroquímica del evento ignimbrítico del Pleistoceno medio (0.33 M.a.) del Valle Central de Costa Rica. Tésis de Licenciatura, Universidad de Costa Rica, 170 pp.

Pichler, H., Weyl, R., 1975. Magmatism and crustal evolution in Costa Rica. Geol. Rundsch. 64, 457–475.

Piedra, J., 1979. Geología del área norte de los Cerros de Escazú, cordillera de Talamanca, Costa Rica. Informe Semestral Ene-Jun, vol. 79, pp. 99–137. Instituto Geográfico Nacional, Costa Rica.

Ponce, D.A., Case, Y.J.E., 1987. Geophysical interpretation of Costa Rica. In: Mineral Resources Assessment of the Republic of Costa Rica. U.S. Geological Surv., Misc. Invest, pp. I-1865, 8–17.

Protti, R., 1986. Geología del flanco sur del volcán Barba. Boletín de vulcanología 17, Ovsicori, UNA.

Protti, M., Schwartz, Y.S., 1994. Mechanics of back arc deformation in Costa Rica: evidence from an aftershock study of the April 22, 1991, Valle de La Estrella, Costa Rica, earthquake ($M_w = 7.7$). Tectonics 13 (5), 1093–1107.

Protti, M., Güendel, F., McNally, K., 1994. The geometry of the Wadati-Benioff zone under southern Central America and its tectonic significance: results from a high-resolution local seismographic network. Phys. Earth Planet. Inter. 84, 271–287. http://dx.doi.org/10.1016/0031-9201(94)90046-9.

Protti, J.M., Schwartz, S.Y., Zandt, G., 1996. Simultaneous inversion for earthquake location and velocity structure beneath central Costa Rica. Bull. Seism. Soc. Am. 86 (1A), 19–31.

Protti, R., 1996a. Evidencias de glaciación en el Valle del general (Costa Rica) durante el Pleistoceno tardío. Revista geológica de Centroamérica 19, 75–85.

Protti, R., 1996b. Monitoreo de desplazamiento de la falla La Garita (Costa Rica) entre marzo y diciembre de 1990. Nota técnica. Revista Centroamericana de Geología 19, 183–185.

Protti-Quesada, J.M., 1994. The Most Recent Large Earthquakes in Costa Rica (1990 Mw 7.0 and 1991 Mw 7.6) and Three-Dimensional Crustal and Upper Mantle P-Wave Velocity Structure of Central Costa Rica. Ph.D. Dissertation, University of California, Santa Cruz, p. 116.

Protti, R., 2013. Megadeslizamiento del flanco norte del volcán Cacahuatique, Morazán, El Salvador. Revista Geológica de América Central 49, 121–127.

Quiesa, S., Civelli, G., Guillot, P.Y., Mora, O., Alvarado, G.E., 1992. Rocas piroclásticas asociadas con la formación de la Caldera de Guayabo, cordillera de Guanacaste, Costa Rica. Rev. Geol. Amér. Central 14, 59–75.

Quintero, R., Segura, J., Jimenez, W., Vega, F., 2004. Evento Principal y réplicas del sismo del 20 de noviembre del 2004 (Mw = 6.4) en Costa Rica. Observatorio Vulcanológico y Sismológico de Costa Rica, OVSICORI-UNA, Costa Rica. Formato pdf. pp. 28.

Rapprich, V., Erban, V., Fárová, K., Kopačková, V., Bellon, H., Hernández, W., 2010. Volcanic history of the Conchagua Peninsula (eastern El Salvador). J. Geosci. 55 (2), 95–112.

Ritmann, A., 1963. Les volcans et leur activité. Masson et Cie, 461.

Rivier, F., 1983. Síntesis geológica y mapa geológico del área del bajo Tempisque, Guanacaste, Costa Rica. Informe Semestral Ene-Jun, vol. 83, pp. 30–70. Instituto Geográfico Nacional Costa Rica.

Rogers, R.D., 2003. Jurassic-Recent Tectonic and Stratigraphic History of the Chortis Block of Honduras and Nicaragua (Northern Central America). Ph.D. Dissertation, The University of Texas at Austin, p. 289.

Roy, J.A., Matthews, S.L., Lachniet, M.S., 2010. Late Quaternary glaciation and equilibrium-line altitudes of the Mayan Ice Cap,Guatemala, Central America. Quatern. Res. 74, 1–7 Elsevier. University of Nevada, Las Vegas, Department of Geoscience, 4505 Maryland Parkway, Las Vegas, NV 89154, USA.

Rubio, A., 1949. Notas sobre la Geología de Panamá. Ministerio de Educación, Panamá.

Ruprecht, P., Terry, P., 2013. Feeding andesitic eruptions with a high-speed connection from the mantle. Nature.

Sadner, G., 1964. La colonización agrícola de Costa Rica. Informe Semestral Ene-Jun, vol. 64, pp. 21–26. Instituto geográfico Nacional Costa Rica.

Saenz, R., 1981. Edades radiométricas de algunas rocas de Costa Rica. Depto. de Geología, MIn. E. M. Costa Rica.

Sanchez, M., 1979. Consideraciones generales sobre la biogeografía y su aplicación a la cuenca inferior y media del río Mosca, Costa Rica. Tesis de Licenciatura en Geografía, UCR Costa Rica.

Sandoval, F., 1971. Geología de la región Noreste del valle Central (Hoja Grecia). Inf. Tecn. y Notas Geol. (Reimp. 9), Dir. Geol. Minas y petrol., Costa Rica, 44 pp.

Sapper, K., 1925. Los volcanes de la América Central. Max Niemeyer, Halle, 144 pp. Deutschland.

Sapper, K., 1943. In: Trejos Quirós, J.F. (Ed.), Viajes a varias partes de la República de Costa Rica 1899-1924. Imprenta Universal, 140 pp. Costa Rica.

Savage, J.M., 1966. The origin and history of the Central American herpetofauna. Copeia, 719-766.

Sayles, R.W., 1931. Bermuda during the Ice Age. Amer. Acad. Arts Sci. 66, 381-468.

Schaufelberger, P., 1935. Un estudio geológico de la meseta Central Occidental. Rev. Inst. de Defensa del Café 1 (2), 148-160.

Schimels, B., 1901. Recent decline in the level of lake Nicaragua. Amer. Geol. 28, 396-398.

Schmidt-Effing, R., 1975. El primer hallazgo de amonitas en América Central Meridional y notas sobre las facies Cretácicas en dicha región. Informe Semestral Ene-Jun, vol. 75, pp. 53-61. Instituto Geográfico Nacional, Costa Rica.

Schmidt-Effing, R., 1980a. Geodynamic history of oceanic crust en Southern Central América. Berliner Geowiss. ABH A19, 201-202.

Schmidt-Effing, R., 1980b. Rasgos fundamentales de la historia del Complejo de Nicoya (América Central meridional). Brenesia 18, 231-252.

Schmincke, H., Kutterolf, S., Perez, W., Rausch, J., Freundt, A., Strauch, W., 2008. Walking through volcanic mud: the 2,100 year-old Acahualinca footprints (Nicaragua). I. Stratigraphy, lithology, volcanology and age of the Acahualinca section. Bull. Volcanol. 51 (5), 479-493.

Schubert, C., 1984. Investigaciones sobre el Cuaternario de la República Dominicana. Instituto Panamericano de Geografía e Historia. revista Geográfica n° 99, pp. 69-92. México DF, México.

Shakun, J.D., Clark, P.U., Feng He, Marcott, S.A., Mix, A.C., Liu, Z., Otto-Bliesner, B., Schmittner, A., Bard, E., 2012. Global warming preceded by increasing carbon dioxide concentrations during the last deglaciation. Nature 484, 49-54.

Siebert, L., Alvarado, G.E., Vallance, J.W., van wyk de Vries, B., 2006. Large-volume volcanic edifice failures in Central America and associated hazards. In: Rose, W.I., Bluth, G.J.S., Carr, M.J., Ewert, J.W., Patino, L.C., Vallance, J.W. (Eds.), Volcanic Hazards in Central America. In: Geol. Soc. Amer. Spec. Pap., vol. 412, pp. 1-26.

Silva Romo, G., Mendoza Rosales, C., 2009. Evaluación geológica de los modelos para El truncameinto cenozóico Del sur de México: Erosión por subducción y detachment del bloque Chortis. Revista mexicana de ciencias geológicas 26 (1), 163-164.

Silver, E.A., Reed, D.L., Tagudin, J.E., Heil, Y.D.J., 1990. Implications of the north and south Panama thrust belt for the origin of the Panama Orocline. Tectonics 9, 261-281.

Simkin, T., Siebert, L., 1994. Volcanoes of the World, second ed. Geoscience Press for the Smithsonian Institution. USA, Tucson.

Sirocko, F., et al., 2005. A late Eemian aridity pulse in central Europe during the last glacial inception. Nature 436, 833-836.

Smithsonian Institution. Global Volcanism Program 2007. Department of Mineral Sciences. National Museum of Natural History.

Sprechmann, P., 1982. Estratigrafía de Costa Rica I – Unidades Estratigráficas sedimentarias V. congreso Lat. de Geol. Actas I-55-71, Argentina.

Sprechmann, P., et al., 1979. Estratigrafía de Costa Rica (informe preliminar). Escuela Centroamericana de GeoL UCR Costa Rica.

Stewart, R.H., 1978. Preliminary geology, el Volcan Region, province of Chiriquı Republic of Panama. Panama Canal Company.

Tarling, M.P., Tarling, D.H., 1973. La derive des continents. Doin Editeurs, Paris.

Taylor, G.D., 1975. The Geology of the Limón Area of Costa Rica. Louisiana State Univ. Dept. of Geol, USA.

Thompson, L.G., 1980. Glaciological investigations of the tropical Quelccaya ice cap, Peru. J. Glaciol. 25 (91), 69–84.

Tournon, J., 1971. Apuntes sobre la geología de la península de Santa Elena. Manuscrito publicado po la Universidad de Costa Rica, 4 pp.

Tournon, J., 1972. Présence de basaltes alcalins récents au Costa Rica (Amérique Centrale). Bull. volcan. XXXVI (1), 140–147.

Tournon, J., 1980. Contribución a la morfología de la parte occidental de la cordillera Central. Informe Semestral Ene-Jun, vol. 80. Instituto Geográfico Nacional, Costa Rica.

Tournon, J., 1983. La cadena volcánica cuaternaria de Costa Rica, Composiciones químicas de las lavas, presencia de dos series. Informe Semestral Ene-Jun, vol. 83. Instituto geográfico Nacional, Costa Rica.

Tournon, J., 1984. Magmatismes du Mesoïque à l'Actuel en Arnérique Centrale L'exemple du Costa Rica, des ophiolites aux andesites. Thèse doctorat d'Etat, Université de Paris VI, 318 pp.

Tournon, J., Alvarado, G., 1995. Mapa Geológico de Costa Rica. Escala1:500.000. más texto. Ministerio de Relaciones Exteriores de Francia, Delegación regional de la cooperación Científica y Tecnica – ICE. Impreso por La Vigie, Dieppe, Francia.

Tricart, J., 1975. Types de lits fluviaux en Amazonie Bresilienne. Ann. Geophys. 473, 1–54.

Tricart, J., 1981. Apercu sur le Quaternaire au Salvador. Bull. Soc. Geol. France 81 (III), 59–68.

Vargas, G., 1978. Diagnóstico y recomendaciones para el manejo y ordenamiento de los recursos naturales en la cuenca del río San Lorenzo, Alajuela, Costa Rica. Tesis de Licenciatura en Geografía UCR, Costa Rica.

Vargas G., 1981. La chaine volcanique de Tilarán et le bassin inférieur du fleuve Bebedero: conditions écologiques, végetation et mise en valeur, Costa Rica. Thèse 3ème Cycle en Géographie, Université de Bordeaux-III, Bordeaux.

Vargas-Ramirez, J.E., 1978. Geología de una parte de la hoja Naranjo. Dir. Geol. Minas y Petrol. Minist. Econ. Ind. y Com Costa Rica.

Weyl, R., 1955a. Vestigios de una glaciación del Pleistoceno en la cordillera de Talamanca, Costa Rica. Informe Trimestral Jul-Sept, vol. 55, pp. 9–32. Instituto Geográfico Nacional, Costa Rica.

Weyl, R., 1955b. Viaje al cerro Chirripó en la cordillera de Talamanca. Informe Trimestral ene-marzo, vol. 55, pp. 7–13. Instituto Geográfico Nacional, Costa Rica.

Weyl, R., 1956. Vulcanismo y plutonismo en el sur de Centro América. Informe Trimestral Jul-Sep, vol. 56, pp. 9–17. Instituto Geográfico Nacional, Costa Rica.

Weyl, R., 1957. Contribución a la Geología de la cordillera de Talamanca de Costa Rica, Centro América. Instituto Geográfico Nacional, Costa Rica.

Weyl, R., 1960. Las ignimbritas centroamericanas. Informe Semestral Ene-Jun, vol. 60, pp. 39–59. Instituto Geográfico Nacional, Costa Rica.

Weyl, R., 1971. La clasificación morfotectónica de Costa Rica Informe Semestral Jul-Dic, vol. 71, pp. 107–125. Instituto Geográfico Nacional, Costa Rica.

Weyl, R., 1980. Geology of Central América. Gebruder Borntraeger, Berlin-Stuttgart, RFA.

Williams, H., 1952. Volcanic history of the Meseta Central Occidental, Costa Rica. Univ. Cal. Pub. Geol. Sci. 29 (4), 145–180.

Williams-Jones, G., Rymer, H., Rothery, D.A., 2003. Gravity changes and passive degassing at the Masaya caldera complex, Nicaragua. J. Volcanol. Geotherm. Res. 123 (1–2), 137–160.

Wilson, J.T., 1959. Geophysics and continental growth. American Scientist 47 (1959), 1–24.

Wilson, J., 1960. Some consequences of expansion of the earth. Nature 185, 880–882.

Wilson, J.T., 1963. A possible origin of the Hawaiian Islands. Can. J. Phys. 41, 863–870.

Winter, A., et al., 2003. Orbital control of low-latitude seasonality during the Eemian. Geophys. Res. Lett. 30 (4), 12.

Woodring, W.P., Malavassi, E., 1961. Miocene foraminifera mollusks and barnacles from the Valle Central, Costa Rica. J. Paleont. 35 (3), 489.

Glossary

A

Advance pre-existing rivers and the formation of a relief. It is said in reference to thalwegs

Adventitious a parasite cone on a volcano whose activity was very low

Alkaline soil whose pH is higher than 7; usually calcareous soils

Alluvium sedimentary deposits formed by pebbles, sands, silts, and clays deposited by the fluvial action

Ammonite fossil; a mollusk that lived in the Jurassic and Cretaceous eras

Amphibolite transformation from metamorphic rock, deep calcareous pelitic rock, or basic eruptive rocks

Andesite lava of intermediate acidity that has silica content between 52% and 60%

Anticline structural form folded in the shape of a roof

B

Barrier Reef coral reef that is remote to the coastline, that sometimes leaves a lake between the coast and the same reef

Basalt basic volcanic rock. It contains less than 52% of silica. It gives extensive fluid flows

Batholith deep magmatic rise, usually of granite or granodiorite, by internal cooling effects. A batholith is the heart of a mountain range

Biostasy period which allows the vegetation to protect soils. A phase of biological stability giving rise to alteration *in situ*

Burning cloud also "nuée ardente," or pyroclastic volcanic flow. A thick, gaseous, volcanic and acidic emission formed from violent eruptions with abundant pumice. Burning clouds descend from craters with temperatures above 700 °C

C

Caldera usually circular, a depression formerly occupied by a volcanic focus (a collapsed caldera occurs when the magmatic chamber of a volcano collapses). Whether it remains part of the volcanic cone depends if it is an explosive, or non-explosive caldera

Cell small depressions that fit the rock by the chemical effect of dissolution

Chipboard heterogeneous set of rocks of volcanic and sedimentary origin, united in a clay matrix

Climax optimal state of a relatively stable balance between vegetation and soil

Colluvial side accumulations of sediment in a valley whose travel occurs by way of gravity, without the medium of water

D

Dacite volcanic rock

Deciduous vegetation that temporarily loses its leaves

Diabase an element/mineral component of volcanic rock

155

Diorite igneous rock formed by plagioclase associated with iron minerals (biotite and hornblende)

Dip slope that has a stratification. You must combine with the address that is measured with a magnetic compass

Dissected eroded, cut, or undermined by physical/meteorological mechanisms

E

Eemian Riss-Würm interglacial period of overheating of the terrestrial globe. It occurred between 160,000 and 80,000 years BP, and was characterized by a sharp rise in the sea level of 140 m

Escarpment rocky, steep slope (greater than 45 °)

Estran (Foreshore) rocky plain eroded by the sea at the base of a cliff

F

Fault fractured surface of Earth's crust due to adjustment of tectonic plate movements, or locally due to stresses on hard surfaces causing the displacement of two compartments. If it is recent, it's a **mirror fault** of failure. If it has been altered by erosion, it's a **fault scarp**. If the scarp has receded by erosive effects, it's a **line fault scarp**. If the block sunk by erosion remains in the upper position, it's a **reverse fault scarp**

Flandrian Coastal Arrow cord of sand and boulders accumulated by the littoral drift. The coastal cords are formed in the same way beaches originate

Flandrian period an optimum climate was reached 6000 years BP, which was marked throughout the world by a 4 m rise in the sea level, due to widespread melting. Also referred to as **Flandrian transgression**

Foehn hot and dry wind descending from a mountain. It's done by **katabatic effect**, due to subsidence, which makes the air heated and lose its relative humidity. Its strength is due to the fact of it being channeled by valleys

Fringe Reef coral reef that hits the beach or coastline

Ftanites volcanic rocks

Fumaroles post-eruptive volcanic phenomenon that translates to emissions of sulfurous fumes and other components with high temperatures. (This occurs at the Poas Volcano.)

G

Gabbro intrusive igneous rock. Its composition is similar to basalt rock

Geyser source of hot water associated with volcanic activity that arises suddenly, expelling into the air

Giant marmite circular cavity undermining the bedrock of a river or the coast, caused by the abrasive movement of boulders

Glacis vast sedimentary area composed of fine particles that extends to the feet of an alluvial fan, and imperceptibly comes into contact with the plain

Graben tectonic pit

Greywacke volcanic deposits characterized by black sands, lapilli, and pumice

H

Halophyte plant living in a salty, aqueous medium

Heterometric sedimentary material in which fine and coarse materials coexist without a scale of values

Hypercalcareous sediment oversaturated in calcium carbonate

Hogback rectilinear crest in a folded sedimentary structure. (The Appalachian Ridge.)

I

Ice sheet glacier of convex shape that covers a large area in Antarctica and Greenland

Igneous comes from the Earth's interior. Igneous rocks can be intrusive or extrusive

Ignimbrite acidic, volcanic deposit that is consolidated after being erupted from stratovolcanoes; associated with violent volcanic eruptions (burning clouds)

Isobath isoline, indicating a depth determined in oceans, seas or lakes

J

Jasper rock in the process of metamorphism of sedimentary origin (clay)

Joints cracks in rocks. Sedimentary rocks are perpendicular or oblique to stratification. In crystalline rocks, joints are curved

K

Karst process of dissolution via the effects of rainwater; mainly of limestone that is rich in calcium and magnesium carbonate

L

Lagoon semi-open seawater surface surrounded by banks of corals

Lahar flowing of mud and assorted materials of volcanic slopes that are associated with a volcanic eruption. When it comes to the overflow of a crater lake, the flow may be hot. In the case of ice suddenly melting in a volcanic cone, the flow can be cold. They can cause great harm

Lamellibranches marine or freshwater mollusks fitted with a bivalve shell

Lapilli volcanic-projection-type slag containing fragments of lava

Latite volcanic rock

Limestone sedimentary rock composed of marine microorganisms

Lumachelles (Coquina) sea mollusks commonly known as porcelain

M

Maar eruptive volcanic gas depression or crater

Madreporic of coral origin

Meander very sinuous curve formed by a river in a plain (lower course) due to lack of competence of the river. The meanders are typical of low tropical coastlines

Mesa aclinal sedimentary structural form which gives rise to an elevated and isolated plateau

Monocline inclined, but not folded, sedimentary structure. If the top layer is resistant and consistent, with respect to the bottom, it gives origin to a **cuesta structure**

Multiconvex modeling first form of the humid, tropical landscapes where the relief, by effects of erosion, forms round hills, or domes

Multifaceted modeling second form of the humid, tropical landscape. The relief forms adopt chevron forms by effects of erosion, in direct accordance with the underlying structure

N

Nappe (French) a huge recumbent fold with both limbs practically horizontal which has been forced a great distance (several miles) over the underlying formations, thus covering them like a cloth

Naze or Ness a promontory or headland

Nuée ardente (French) or Pelean cloud the blast of hot highly gas-charged, swiftly moving lava fragments sometimes ejected horizontally from a volcano when upward movement it has been obstructed

O

Ofilitic volcanic rock complex formed by intrusive, basic, and ultra basic rocks (diorites, dolerites, gabbros, and peridotites), from seabed (greenstones), typical of geosynclines volcanism

Olive small, porcelain-type of dark green mollusk

Ombrothermic dense, tropical rainforest with elevated temperatures
Orogenesis also tectogenesis. The result of the lifting relief process (i.e., the Andes Range)
Ox bow lake formed in an abandoned meander

P

Peridotite olive-green plutonic rock composed of olivine
Piedmont section in contact with a mountain massif characterized by the presence of alluvial fans
Pillow lavas basalts cooled in an aqueous medium, taking the shape of pillows
Puddingue Chipboard River. Strong cohesive composed of boulders of different sizes, generally cemented by a clay matrix

R

Radiolarites sedimentary rocks composed of organic marine sediments—the radiolarians
Rasa level of marine erosion on a rocky surface
Rhesistasic driest period linking the evolution of soils to plant cover, and forms of modeling. It offers explanation for sedimentary cycles. The opposite is a **biostasic period**
Rhyolite volcanic rock

S

Shale metamorphic rock of sedimentary origin (compacted clays), i.e., flysch
Shield volcano depressed volcanic cone in its center. The slopes of the volcano are shallow. Water converges towards the ancient volcanic focus
Solifluction landslides of clays and other materials due to massive contributions of floods. It may cause laminar landslides, or mass catastrophic landslides and mud flows
Subduction area of confrontation of two tectonic plates, where the first plunges into the mantle
Subsidence progressive collapse of a basin, or a tectonic pit, beneath the weight of the sedimentation that is deposited
Superimposition thalwegs of rivers that have subsequently fitted an ancient, pre-existing structural surface
Syncline folded, canoe-shaped sedimentary form

T

Table said of a liquid squeezed between two strata (phreatic mantle)
Thalwegs maximum depth line in the longitudinal profile of a river
Tarn lake of glacial origin
Toleitic type of nether rock composed of phenocrysts of hornblende, andesite, biotite, muscovite, and sometimes garnet
Trade winds (s) the NE wind of the tropical zone

U

Upwelling rocks uncovered, or naked, in good condition

W

Würm glacial period corresponding to the fourth European glaciation, extending from 90,000 to 12,000 years BP. In the United States, it corresponds to the Illinoian-Wisconsinian Stages

X

Xerophyte said of vegetation, or animals accustomed to the lack of water, i.e., prolonged dry seasons

Index

159